# Sustainable Development, Leadership, and Innovations

# Sustainable Development, Leadership, and Innovations

Dalia Štreimikienė,
Asta Mikalauskiene,
and Remigijus Ciegis

CRC Press is an imprint of the
Taylor & Francis Group, an **informa** business

CRC Press
Taylor & Francis Group
6000 Broken Sound Parkway NW, Suite 300
Boca Raton, FL 33487-2742



**CRC Press is an imprint of Taylor & Francis Group, an Informa business**



Printed on acid-free paper

International Standard Book Number-13 978-0-367-36943-9 (Hardback)



**Visit the Taylor & Francis Web site at www.taylorandfrancis.com and the CRC Press Web site at www.crcpress.com**

# Contents

# Authors

**Prof. Dr. Dalia Štreimikienė** is Professor at Vilnius University, Kaunas, and Leading Research Associate at Lithuanian Energy Institute. Her main areas of research are sustainable development, sustainability assessment, and corporate social responsibility. She is the author of more than 100 papers in international journals indexed at WoS. H index 19. She is the invited editor for several special issues of the journal *Sustainability* and Editor-in-Chief of the international journal *Transformations in Business and Economics*, indexed at WoS. Orcid: 0000-0002-3247-9912. www.researchgate.net/profile/Dalia_Streimikiene2.

**Prof. Dr. Asta Mikalauskiene** is Professor at Vilnius University, Kaunas. Her main areas of research are sustainable development, sustainability assessment, and corporate social responsibility. She is the author of many papers in international journals indexed at WoS. Orcid: 0000-0002-4301-2058.
www.researchgate.net/profile/Asta_Mikalauskiene.

**Prof. Habil. Dr. Remigijus Ciegis** is Professor at Vilnius University, Kaunas. His main areas of research are sustainable development, sustainability assessment, and environmental regulation. He is the author of more than 20 scientific monographs on sustainable development. Orcid: 0000-0002-3538-7378.

# Preface

In this book, the significance of sustainable leadership is revealed in order to encourage sustainable development of Lithuania. In addition, on the basis of research insights along with practical achievements and success stories of leaders, the connections among sustainable development, corporate social responsibility, and sustainable leadership, as well as the causal chain, which perfectly discloses the synergistic effect and power of these conceptions, are clearly presented in a way that is understandable to everyone.

In the book, sustainable leadership is viewed as the most significant means for the creation of sustainable organizational culture, while the formulated principles of sustainable leadership are universal and suitable for implementing sustainability in organizations of both private and public sectors.

The case studies of sustainable leadership were selected from Lithuania. Here, as in many other countries around the world, sustainable development is the main economic and social direction of the country. Only sustainable development, based on the compatibility of economic, social, and environmental objectives, can guarantee the social welfare and happiness of residents, as well as preserve the environment for present and future generations. However, Lithuania, as a new member of the European Union (EU) and a transition economy, is facing many changes and Lithuanian case studies provide good examples for other transition countries. Although many strategies and development plans have been written in our country, most of the objectives of social development have still not been achieved. Here, major inequality among the people exists alongside high poverty indicators, and the indicators of income, health, and happiness are among the worst in the EU. The weakened psychological resilience of people and high suicide rates, indicating a high level of public frustration and insecurity, are especially worrisome.

That said, the corporate social responsibility, by employing sustainable leadership, could solve most of the social problems in Lithuania and other transition countries encountering similar problems, such as poverty, income inequality, social exclusion, and the accompanying disappointment and depression, as well as other health and spiritual problems. The solution to these problems would also provide more happiness to the country's residents and would reduce the rate of suicides, which is currently one of the highest in the world. The issues of corporate social responsibility, sustainable business, and sustainable leadership are extremely relevant to all transition countries.

# Acknowledgment

Funding from the European Social Fund (Project No. 09.3.3-LMT-K-712-14-0053) under grant agreement with the Lithuanian Council of Science (LMTLT) was used for the preparation of the second chapter of the monograph.

# 1 Sustainable Development and Organizational Sustainability

## 1.1 THE CONCEPT OF SUSTAINABLE DEVELOPMENT AND THE MAIN DIMENSIONS OF SUSTAINABILITY

The concept of sustainable development includes three closely related areas: economy, social environment, and natural environment. All these areas are inseparable in order to encourage sustainable development. Over the last decades, the problems of the loss of natural resources and global environmental pollution have risen sharply. Dwindling, rare, and expensive natural resources, acidification of freshwater bodies, soil erosion, and the ozone depletion, as well as global warming and natural cataclysms related to it—these particularly sensitive global environmental problems have encouraged world governments to implement sustainable development policies and businesses to introduce socially responsible and sustainable business ideas in their activities (Raworth 2012; Nykvist et al. 2013). In addition to global environmental problems, there are a number of sensitive social issues in the world such as famine, poverty, high morbidity and mortality rates, social inequality, and the constant increase in social tensions (Tepperman and Curtis 2003). These issues are closely linked to the increased risk of terrorism and the threat of dictatorial regimes to peace and security around the world. The concept of sustainable development plays an essential role in today's unstable world as its objectives are in line with the development interests of all countries and promote cooperation between countries in order to ensure the sustainable future of humanity.

The concept of sustainable development was formalized in 1987, following the publication of the report by the United Nations (UN) World Commission on Environment and Development named *Our Common Future* (1987). For the first time, in this report, the term sustainable development was defined and a new qualitative definition of economic growth—rapid and at the same time socially and environmentally sustainable economic growth—was formulated. Thus, sustainable development includes three dimensions: economic, social, and environmental. The concept of sustainable development states that all economic growth must take into account the economic and social aspects and consequences accompanying it (Dearing et al. 2014).

The society and its established systems function in a closed ecosystem and use limited resources, such as water, soil, and biomass; therefore, they must absorb the pollution caused by public activities in the ecosystem (United Nations 2014). Rapid population growth and rapid economic growth, which are related to the increasing environmental pollution and the loss of natural resources, have led to the growing concerns of scientists and the public primarily over environmental issues and, subsequently, over further opportunities for societal development and the achievement of social objectives of sustainable development. The polluted environment and the loss of natural resources have sent the first signals to society of irreversible processes in the world related to economic development; these processes not only may determine the lifestyle of our generation but also may have even greater negative consequences for future generations (Rockström et al. 2009). Therefore, in 1987, in the report *Our Common Future* (World Commission on Environment and Development, 1987), a vision of future development—*sustainable development*—was formulated.

According to the UN definition, sustainable development *is a type of development that meets the present economic, social and environmental needs of society without compromising the ability of future generations to meet their own needs.* Currently, a clear contrast between developed and developing countries exists, and a direct link between poverty and environmental problems has been determined (Steffen et al. 2015). It has allowed us to extend the content of sustainability together with the concept of sustainable development. In 1992, at the UN Conference on Environment and Development in Rio de Janeiro, the main sustainable development provisions were made and an action program of sustainable development implementation—Agenda 21—was approved.

Thus, fundamentally sustainable development provides a better quality of life for our generation as well as those in the future (Elliott 2005). Quality of life includes such important aspects as the environmental conditions, state of personal and public health, security, material welfare, psychological well-being, emotional condition, social ties, etc. Moreover, the improvement of the environmental condition is not an end in itself, as environmental imbalance sooner or later disturbs economic development and deteriorates people's quality of life. Therefore, the main objective of sustainable development is to meet the people's basic physical, material, social, and spiritual needs and to allow them to freely choose, develop, and use their potential (Ciegis 2004). It is possible to achieve it if there is an effective environmental protection and healthcare system, accessible education and information; also, if people have conditions to work and earn money, participate in public activities, etc. Of course, in order to meet the present and future basic needs of citizens and to ensure a high quality of life, it is necessary to strengthen the economy (Ciegis, Ramanauskiene, and Martinkus 2009). However, economic growth and social welfare alone are not enough to make people happy. Society requires opportunities for freedom, choice, self-realization, and spiritual growth. Sustainable development is often equated with such environmental activities as waste sorting, recycling, biodiversity protection, etc. Nevertheless, the concept of sustainable development is much broader and essential aspects of human social development and improvement should not be forgotten.

The World Bank defines sustainable development very succinctly, as a development that continues. In 1992, in the Rio de Janeiro Declaration on Environment and Development, sustainable development was described as a long-term continuous "development that meets the needs of the present without compromising the ability of future generations to meet their own needs" (Keating 2004). These two definitions reveal that the World Bank defines sustainable development only as a continuous process, while the Rio de Janeiro Declaration emphasizes not environmental or economic but continuous societal development, which can be achieved by employing three essential and equivalent components (environmental protection, economic growth, and social development).

Sustainable development is often described as the result of growing awareness of the common interface between environmental problems and socio-economic issues related to poverty, inequality, and concerns about healthy future humanity (Hopwood, Mellor, and O'Brien 2005; Mauerhofer 2008). According to this definition, sustainable development is understood as the result of the successful solution of environmental, social, and economic problems. In modern times, sustainable development is first perceived as the desire of humanity to stimulate a coordinated development all around the world, in every country, and in all areas of social life, that brings about harmonious and minimized harm to the human and environment as well as minimizes social opposition. Sustainable development is based not only on a high level of environmental protection and environmental quality improvement but also on the competitive social market economy and an effort to achieve significant social progress (United Nations General Assembly 2012). Although sustainable development can be interpreted in different ways, the main point stays the same: it is a never-ending process seeking a compromise among environmental, economic, and social objectives of society, and enabling general welfare for present and future generations within the limits of permissible impact on the environment (World Business Council for Sustainable Development 2014).

According to the Academic Dictionary of Lithuanian, the word "develop" means to grow, mature, become stronger and more complicated than before; it includes both quantitative and qualitative changes. Meanwhile, in terms of development in the context of sustainable development, one of the most frequently discussed issues is whether economic growth and sustainable development are compatible and whether the growth can be seen as a part of the sustainable development process. In the report of G. H. Brundtland (*Our Common Future*), economic growth is viewed as an indispensable condition of development and it is indicated that only economic growth can ensure a successful solution to poverty and environmental pollution problems. In the report, recommendations are even provided for maintaining high rates of economic growth. Developing countries should maintain stable economic growth rates of 5–6 percent, while developed countries should "slow down" their economic growth by halving current economic growth rates ( World Commission on Environment and Development, 1987).

Though the problems and main purposes of sustainable development vary in different countries and regions, the concept of sustainable development still includes three essential equivalent components: environmental protection, economic development, and social development (Spangenberg 2002a, 2002b). Only such development can guarantee security for people and make them happy. Although the concept of sustainable development usually refers to the three main components, one element that is not directly mentioned can also be distinguished: the institutional one; this element includes formalization, validation, maintenance, and support of solutions of economic, environmental, and social problems as well as the creation and adaptation of infrastructure, etc. Institutions play an important role in the implementation of economic, social, and environmental objectives of sustainable development. The high quality of institutions and the developed institutional capital allow countries to achieve their sustainable development aims more successfully. Successful compatibility of economic, environmental, and social objectives can only be achieved by creating a favorable institutional environment, as the institutions and their decisions influence economic, social, and environmental dimensions, whereas these dimensions have an impact on the institutions (Spangenberg 2002a, 2002b; Hajer 2011). It can be said that sustainable development is not possible without the participation of institutions, not only in solving the specific, identified challenges but also in making decisions at various levels that are important for sustainable development.

Notwithstanding that all the components of sustainable development and their enveloping institutional dimension are united by the overall purpose of sustainable development, each component has its own objectives and specific functions (Pattberg 2012). Economic sustainability primarily focuses on material needs and their fulfilment since the economic approach of sustainability covers the requirements of sufficient and stable economic growth, such as preserving financial stability, low and steady inflation rates, the ability to invest, and innovations; in addition, it requires the integration of economic activity and productivity of ecosystems. Thus, economic sustainability not only is business oriented, which should provide long-term benefits that contribute to people's financial welfare, but also must take into account the efficient and responsible use of natural resources (Mikalauskiene and Streimikiene 2014).

It is obvious that economic growth has an impact not only on the environment but also on the social development of society. The positive consequences of social economy development are the following: the improvement of living standards, income redistribution, poverty reduction; the negative ones are: unfavorable lifestyle changes, private costs, and social costs. Therefore, the economic dimension of sustainable development includes such development which creates conditions for long-term, stable economic development by using available resources more efficiently.

Environmental aspects of sustainable development include pollution control, cleaner and sustainable production, eco-design, waste reduction, etc. The environmental aspect of sustainability is mainly focused on integrity, productivity, and preservation of the stability of biological and physical systems (Ciegis 2004, 2006). The environmental sustainability first seeks to ensure the protection of the environment so that natural resources are preserved and accessible for future generations. The environmental sustainability is based on the fact that people can exhaust or exploit too many natural resources without leaving anything for future generations, only contaminated water, impoverished soil, and cut-down forests. Hence, the environmental dimension of sustainable development is directed at development in which the environmental and natural resources are used wisely and effectively, in this way preserving them for future generations.

Sustainable development is impossible without social sustainability, which means a free, democratic, healthy, secure, conscious, and righteous society based on social inclusion and cohesion and which respects fundamental rights and cultural diversity, as well as ensuring equal opportunities and combating all forms of discrimination. Thus, social sustainability promotes social development, justice, and at the same time seeks to strengthen corporate social responsibility. In addition, the role of local communities, their knowledge, experience, traditions, and participation in order to achieve sustainable development, is also very important. The social aspect of sustainable development can be analyzed as the development of human capital, which is related to the entirety of human possibilities, but not to the interaction among people and the rules they follow that are described by the before-mentioned institutional dimension of sustainable development or institutional capital (Elzen, Geels, and Green 2005).

If the components of sustainable development do not cooperate, the enveloping institutional dimension of all these components, which aims to achieve the best-balanced results in economic, environmental, and social aspects, has not been successfully achieved. Therefore, every component of sustainable development is important; however, in order to understand the process of sustainable development, the best option is to analyze not the individual parts of the process but their interaction. In addition, when analyzing the process of sustainable development, one should pay particular attention to the institutional dimension since only effective and well-functioning institutions can provide the basis for sustainable development (Pfahl 2005). In the context of sustainable development, institutions include the legal structure, formal and informal markets, and various public bodies, as well as interpersonal networks, rules, and norms that are followed by members of society. The institutional dimension not only influences other dimensions and their coordination but also connects them into one entirety.

Thus, the implementation of sustainable development requires high-quality institutions to ensure good governance, which can be associated with the best possible decision-making and the process of its successful implementation at all levels. Good governance of sustainable development should be displayed in national, regional, and organizational levels because, in order to implement the objectives of sustainable development, effective, transparent, and democratic institutions are necessary at all levels (United Nations 2018). Good governance is impossible without public participation. It should be emphasized that the basis of implementation of sustainable development in each country is an appropriate environmental, social, and economic policy that is impossible without the broad public participation not only in addressing specific problems but also in making important decisions for sustainable development at various levels. This policy should be effective and send appropriate signals to market entities (business, residents) primarily promoting clean and sustainable production together with sustainable and responsible consumption (United Nations 2018).

In summary, although there are different interpretations of the concept of sustainable development and various explanations of the dimensions of sustainable development, the content of this concept remains the same: to seek economic, environmental, and social sustainability. Management of sustainable development is a complex process consisting of such elements as the number of organizations, political instruments, and financial mechanisms, as well as rules, procedures, and norms that regulate sustainable development processes at all levels.

The majority of EU and other developed countries pay increasingly more attention to sustainable development as well as sustainable and socially responsible business development. Strategic documents of countries' sustainable development, including the sustainable development of transport, industry, agriculture, and other areas, are prepared or included in other programs, where the emphasis is on environmental aspects and economic and social aspects which are associated with them. Many governments all over the world have established a variety of institutions to help the industries implement sustainable development measures and raise their qualification in this area—for example, setting up cleaner production centres or special departments providing technical assistance for companies interested in green businesses. The activities of these departments aim to build the trust of companies and ensure that they receive all the necessary information related to the installation of cleaner production measures and their benefits. Increasingly, more attention is being paid to scientific and technological progress as well as innovations; appropriate programs have been created and specific support measures have been applied, while the entities are encouraged to create and use new, effective, resource and energy-efficient technologies which are neutral in terms of pollution; they are financially supported and other measures of encouragement are applied, such as social marketing, dissemination of information, publicity, various awards, etc.

In Lithuania, environmental legislation is practically in line with the EU standards and encourages the application of sustainable industrial development measures: developed technical potential, especially in the area of cleaner production, and favorable environment for the implementation of quality and environmental management systems, etc. (National Strategy for Sustainable Development 2009). However, companies still lack knowledge in eco-design, and the resources to develop sustainable innovations independently, whereas society lacks accessible information on the environmental impact of businesses. Due to low public awareness and limited purchasing power, there is still insufficient demand for products and services that are safer for the environment in Lithuania. However, in Lithuania, there are opportunities to optimize the economic and environmental capabilities of the industry by implementing cost-effective preventive measures to address environmental issues. The main trends of sustainable business development in Lithuania are: cleaner production and eco-efficiency (a more rational use of energy and natural resources, minimization of waste and pollutants at source, and recycling); the implementation of quality environmental protection and social management systems involving all employees of the organization; product or service-related sustainable industrial development measures that help to reduce the environmental impact of a product throughout its whole life-cycle: eco-design, producer responsibility, eco-labeling, etc. (National Strategy for Sustainable Development 2009). Sustainable development efficiency reports allow companies to carefully analyze and improve their activities, while their publicity increases public awareness and awareness of sustainable development (Lithuanian Progress Report 2015). However, in order to apply effective measures of sustainable development, it is necessary to create conditions that would encourage businesses to implement these measures and to support them in their implementation.

After the economic crisis, the Lithuanian economy is developing rapidly, but serious problems exist in the social area of sustainable development. In spring 2016, the European Commission announced the macroeconomic assessments of EU Member States, which forecast a decrease in general government deficit, higher investment growth, and reduced level of unemployment in Lithuania. The spring assessments of the European Commission show that the Lithuanian economy grows steadily and in the near future the main driving force of this growth will be not only the increased household consumption but also the recovering export. However, some serious problems are present in Lithuania: the increasing differentiation of income of the Lithuanian population as well as social inequality, poverty, and exclusion. In rural areas of Lithuania, nearly one-third of people live below the poverty line. High emigration rates and depopulation are the gravest problems in Lithuania. As the population decreases, the number of retired people is increasing rapidly, which poses issues for SODRA (the State Social Insurance Fund Board under the Ministry of Social Security and Labor), as well as labor force shortages. Wealth inequality has a strong impact on human health, behavior, beliefs, and values; in addition, it causes social tensions in Lithuania. Horizontal differentiation among society is still growing in Lithuania. An enormous problem is the number of young people who are neither studying nor working. This situation can cause a social explosion. It is clear that these tendencies can only be changed by using innovative educational models. Thus, teachers should foster values of the younger generation, so that they would see work as a value and not just as a source of income.

Lithuania is characterized by high unemployment and low living standards. At the same time, the country has a very low birth rate and high emigration numbers; therefore, sparsely populated regions called “demographic deserts” are forming in Lithuania. In such deserts, there are almost no residents. Basically, the whole of northeastern Lithuania is a sparsely populated region. In such areas in Lithuania, up to ten people live in a square kilometre. This situation exists in 183 municipalities, which represent 45 percent of Lithuania’s area. In the last 20 years, Lithuania lost 20 percent of its population (Lithuanian Department of Statistics 2017). Moreover, this closed circle is developing: due to the declining population, the entire economic and social infrastructure as well as other types of infrastructure are disappearing from the “demographic deserts.” In this situation, the role of socially responsible or sustainable business is essential. This type of business or investments can help to effectively solve these social problems in Lithuania because social support, provided by the country, cannot cope with these issues and guarantee social welfare for the Lithuanian people. The country can prepare policy measures and provide conditions that encourage businesses to be socially responsible and sustainable; however, it is not fit to develop production, to provide work for all the people, or to create government monopoly in individual industries. Only the market economy and competition, rather than a big government monopoly in the economic sectors, can guarantee stable economic growth and a high quality of life for the Lithuanian people.

## 1.2 SUSTAINABLE DEVELOPMENT AND HAPPINESS

Sustainable development includes many economic, social, and environmental aspects that are significant (Griggs et al. 2013). However, if one had to choose only one indicator showing the progress of sustainable development of a country, it would be difficult to decide what indicator it should be. Traditionally, in global statistics, GDP per capita is considered the most important indicator of a country's achieved economic welfare; however, this indicator does not reflect all the important dimensions of sustainable development (Mori and Christodoulou 2012). The main advantage of this indicator is its calculation, which avoids many inaccuracies, artificial assumptions, and biases that are common to other welfare indicators, such as indices of happiness or different indicators and measurements of people's life satisfaction (Mori and Christodoulou 2012).

Nevertheless, various happiness indicators created by scientists are essential in order to determine the priorities of a country's sustainable development, since as the economy grows, the growth of the population's social welfare, in many cases, is not so rapid, and because of stress and tension, the number of people suffering from depression and other forms of personality disorders quickly increases. It is perfectly illustrated by high suicide rates in Lithuania that reflect the disappointment and dissatisfaction of life as well as a wish to take your own life.

As Maslow's theory states, if human needs are more or less satisfied, one feels more or less happy (Tezcan Uysal, Aydemir, and Genç 2017). Therefore, a suitable economic performance indicator could be human happiness. In many countries, human happiness indices are counted by conducting surveys asking if people are happy. Thus, given low GDP per capita levels, index of happiness rapidly increases with boosting incomes; however, if a person makes more than US$10,000 per capita, this dependence is invalid. Hence, given a low GDP per capita level, which is currently common in developing countries, economic growth is very important in order to ensure people's welfare; nevertheless, in developed industrial countries with the highest rates of income, people's happiness does not depend on the growth of GDP per capita. Lithuania has exceeded the US$10,000 per capita threshold and, as a result, the increase in the happiness index is no longer associated with the growth of income.

For a very long time, scientists have had a keen interest in the specific sources of human happiness. Traditionally, it has been considered that human happiness depends on the consumption of goods and services, and as this consumption grows, happiness increases as well. One of the common methods to determine what impacts people's physical and mental health is to test their genetic origins that have been determined by the conditions of biological evolution. In these studies, scientists highlighted physical and psychological needs, which were formed by biological and cultural evolutions, respectively, and the ways to satisfy them (Helliwell, Layard and Sachs 2018).

Ninteenth-century economists believed that happiness, which they called "usefulness," could be measured. In 1950, this belief was denied by neoclassical economists. Welfare economists continued to use the concept of usefulness, by giving it the meaning of happiness; however, they did not try to measure it. It was stated that the levels of usefulness or happiness could not be measured and applicably used for the comparison of these indicators between individual groups or separate individuals (Cipresso, Serino and Riva 2014). This saying was common among the neoclassical economists: we cannot say if person A is happier or has greater usefulness than person B. However, in the last few decades, economists started analyzing people's feelings and identifying what makes people happy and what the causes of this feeling are.

The first question which needs to be answered is whether happiness can be measured as it is a state of mind and feelings. For this reason, psychologists determine the level of happiness by questioning individuals about their feelings. A typical question in surveys analyzing happiness and its factors is: "Considering all the circumstances, how would you describe your state: very happy, happy, unhappy or very unhappy?"

An alternative for such a survey is a questionnaire in which an individual is asked to evaluate his life satisfaction on a scale from 1 to 10, with 1 indicating complete dissatisfaction and 10 complete satisfaction. People answering these questions are not entirely sure if they are correctly measuring their satisfaction in terms of figures; however, it is essential to make sure that people who feel happier would get a higher score. One of the ways to correctly evaluate such surveys is to question the same people at different periods of time. If an individual's situation has not changed considerably since the first survey was carried out, the evaluation of happiness of the individual should not vary dramatically from the previous one. Another way of checking the surveys is to determine how a person's evaluation of happiness correlates with other unbiased indicators of happiness.

It is determined that, in comparison to an average person, individuals who evaluate their happiness with very high scores are:

- considered to be happy people by friends or family members;
- optimists in terms of their own future and the future of other people;
- less likely to commit suicide;
- likely to name more positive than negative things in their life;
- likely to smile more often while interacting;
- healthier.

Such surveys have been conducted in many countries, but the problem is that the word "happiness" has a different meaning in different languages and cultures. Thus, countries have been divided by applying three different approaches: asking people how happy they are, how satisfied they are, and how they would evaluate their life from the worst possible life to the best imaginable life. When applying these three approaches, identical answers have been received in practically all countries.

In recent years, it has been identified that the feelings that people have indicated in these surveys correspond to impartial alterations in their brain that can be measured. Positive feelings correspond to the work of the left hemisphere of the brain, and negative feelings correspond to the work of the right hemisphere of the brain, if the person is right-handed. Psychologists believe that the level of happiness can be determined by comparing it with the level of another individual's happiness, which is achieved by asking adequate questions (White 2014; Anand 2016). This method allows the happiness factors of an individual to be determined, by contrasting the variations of resident happiness evaluations with variations in genetic data and life circumstances. Many studies had been carried out in this area and all of them have had similar results. The first observation of these studies was that the individuals differed genetically due to their attitudes of wanting to be happy. This was the result of daily observations confirmed by experiments and surveys. Some people are happy basically because of their nature, and others, vice versa, are unhappy due to it. The research on factors that influence happiness is based on the answers of individuals about their happiness and life circumstances, and on the answers of random people. The point of this research is to determine the connection between the evaluation of happiness based on numbers and other attributes. Using appropriate statistical methods, a connection between the evaluation of happiness based on numbers and the evaluation of individual attribute based on numbers is made, which later allows the influence of other attributes on happiness to be controlled. However, only those attributes that have been evaluated during the surveys can be controlled. Genetic structure is not evaluated in these surveys; therefore, it cannot be controlled. Nevertheless, it does not prevent the usage of these surveys to research how life circumstances affect human happiness as the genetic structure has more impact on the common sense of happiness than on its response to particular circumstances of life. Therefore, individual A will be happier than individual B in particular life circumstances but this does not mean that a greater number of attributes will make individual A happier than individual B.

The surveys that evaluate the factors of happiness have included such questions as the evaluation of happiness or life satisfaction of individuals, their age, physical health, marital status, labor market situation, education, and income (Anand 2016). Studies have revealed that the impact of unemployment on the happiness index, even with no change in income, is three times bigger than the impact of family income reduction. Even the feeling that your job is not safe reduces happiness more than a drop in the family income. Moreover, the rise of overall unemployment to 10 percent has the same effect on happiness. Meanwhile, the rise of inflation up to 10 percent has a lesser impact on happiness than the reduction of income, the rise of unemployment in the country, or a job loss. A situation in the family has a very strong influence on the individual's evaluation of happiness. A divorced person's happiness decreases by 2.5 times compared to a married one, while the happiness of a person who lives alone decreases by 4.5 times compared to a married person, and the happiness of a widower is two times lower when all the attributes are permanent.

Health is also a very important criterion, as the happiness index decreases three times when the evaluation of health on a scale from 1 to 5 decreases by one point. Therefore, the consumption and variations of income influence happiness but not as much as the other attributes of happiness, such as marital status, job, and health.

If we analyze the data on individuals in a separate country and at a specific period of time, it is obvious that an individual's happiness decreases when income drops (Diener and Oishi 2000). However, a significant number of studies reveal that, with the growth of income, the growth of happiness, which is related to the growth of income, decreases. Here the same law of diminishing marginal utility is observed;

when the saturation point is reached, every additional dollar provides decreasingly fewer benefits or happiness to a person. The same law affects both individuals and countries, which we discussed earlier, as with the growth of GDP per capita, happiness significantly increases only at a low GDP per capita level. An increase in income of $500 brings much more happiness for an individual whose income is lower than $5,000 per year than for an individual whose income is $20,000 per year.

In addition, when comparing GDP per capita and the happiness index in different countries, it can be stated that the happiness index with the rising income per capita increases with declining speed. In 1946–1996, in the USA, with the rapid increase in GDP per capita, the number of residents who considered themselves happy declined. Therefore, there is no connection between the average income per capita and national happiness (Easterlin, McVey, Switek, Sawangfa, and Zweig 2011). In many developed countries, during the last decade, GDP per capita has increased, while the happiness index has remained stable. Thus, a presumption can be made that human happiness is associated with income in a rather strange way: when a person gets richer, his happiness increases; however, within a year, his happiness returns to the same level as before he became richer. Meanwhile, if a person's wealth decreases, he has a difficult time comprehending the deterioration of life quality: with the loss of money, happiness decreases, and it is not restored to the previous level. In other words, people tend to get used to good things quickly, but they undergo poverty more painfully and have difficulties adapting to it. When making economic decisions, people try not to maximize the benefits, but to minimize the loss (Easterlin et al. 2011), even though it seems like a paradox. If we monitor individuals or countries at a particular moment in time, it can be noticed that income brings greater happiness; although, with an increase in income, happiness grows more slowly. On the other hand, when analyzing the economic growth of a rich country within a particular time scale, it can be observed that the growth of GDP per capita does not increase the level of happiness. If a higher income makes people happier, the individuals who get moderately larger incomes over time should be moderately happier in that country. But this is not the case.

This tendency is common in almost all rich countries. In both 1975 and 1998, the majority of rich people considered themselves to be very happy and the majority of poor people considered themselves to be very unhappy. At a specific moment in time, people on higher incomes are happier. Although, over the last 20 years, the incomes of both the rich and the poor have increased by more than 2 percent per year, the percentage of people in both groups that distinguished themselves as "Very happy," "Pretty happy," and "Very unhappy" practically remained unchanged during this period of time. Thus, although the average income increased substantially over this period, the average level of happiness in the country remained the same. This is due to the fact that the happiness of individuals depends on what they want and experience, whereas their desires depend on what they experience and what they see others experiencing. Desires of a human are shaped by adaptation, habits, and competition with other individuals. Because of adaptation, the increased income allows the person to reach a higher level of consumption, i.e., to acquire more expensive and fashionable clothes and other material goods. At first, it gives an individual great pleasure; however, after a while, they get used to it and their satisfaction returns to the primitive level. If income and consumption constantly increase over time, a routine is established and consistently growing income and consumption becomes a norm that does not bring more happiness.

Primarily, the competition includes self-comparison with others. In 2003, R. Layard described the results of a study during which students were interviewed in order to determine their desires. They were asked to choose between two situations: would they choose to earn $50,000 per year while their friends would get two times smaller incomes in comparison to theirs, or would they choose to earn $100,000 per year while the incomes of their friends would increase twice as much as theirs. Most of the students chose the first option. If the incomes and consumption of the individual's friends increase in a similar manner to the individual's income and consumption, the growth of the income of this individual has little influence on the growth of his happiness (Layard 2003). Thus, adaptation and competition answer the question why, at some point in life, richer individuals are happier, but over time and with increased incomes, the happiness index does not get higher in the country.

Therefore, the economic growth cannot guarantee the growth of happiness when the country has reached a certain GDP per capita or a specific level of economic development; nevertheless, neoclassical economists ignored this statement (Layard 2003). Accordingly, we believe that attention should be paid not to the problems of economic growth, but to environmental problems, conservation of natural resources, and social issues.

Another important problem that is raised when analyzing the connection between incomes and happiness is the uneven distribution of poverty and income. The definition of poverty, basic needs, and contingent needs depends on the connection between poverty and income inequality perception. In the eighteenth century, A. Smith stated that he considered necessity to be not only the goods that are necessary for life, but also what a citizen, even with the lowest income, considers as necessary things for a dignified life. J. M. Keynes also identified absolute and relative needs. The relative needs exist to satisfy a human's desire to be superior, which can be considered as insatiable.

According to the statistical data, economic growth does not reduce inequality among countries. Moreover, it even increases the absolute inequality because, if the economy of rich and poor countries continues to grow at the same pace, absolute inequality will increase. The reduction of inequality among countries requires different growth rates, as the economy of poor countries grows faster. Thus, modern economic growth conditions the growth of continuing inequality between countries; nevertheless, while the growth of developing countries results in the rise from poverty of the poor countries, the fact that rich countries continue to get richer, as the absolute gap between countries increases, is not that important. However, two problems arise. First, an additional dollar of income increases the happiness of poor people more than it would the happiness of rich people; therefore, policies must be primarily focused on the income growth of poor people. Second, the phenomenon of competition means that inequality itself is the source of misery. Thus, if economic activities are treated like tools of human happiness which satisfy people's desires and needs, then the problems of inequality and poverty should be tackled first, as the economic growth of poor countries is not enough to secure human happiness; hence, policies should be implemented to reduce inequality among countries.

Interestingly, subjective life satisfaction may depend even on the mood of the analyzed person. The research showed that daily life events may affect mood considerably. Thus, by changing an individual's mood, life events may directly affect the quality of life. On the other hand, it may be stated that it is not the events but the interpretation of them that affects the quality of life. The results of the analysis of life satisfaction and the socio-demographic index are quite interesting. It is revealed that the human life satisfaction level remains quite similar during all stages of adult life. In some countries, for instance, in Malaysia, women's life satisfaction varies depending on their age. Minnesota is one of the happiest states in the USA, where the connection between life satisfaction and age is expressed as a U-shaped curve. Life satisfaction is affected by such important social factors as marital status, employment status, job loss, etc. People who describe themselves as happy usually are sociable and can be seen as extroverts and optimists. Studies reveal that happier individuals live a more environmentally friendly life. Other examples of subjective indicators of life satisfaction may be a job and wage that provides pleasure, as well as satisfaction in marriage and family life.

In all countries where such studies have been conducted, it has been determined that happy and satisfied people are proud of their job, take care of their healthy lifestyle, have a good relationship with their family, have many friends, are good at stress management and devote a lot of time to leisure. It has also been revealed that people feel happier when they achieve their objectives; in other words, goal achievement makes people happy. Positive emotions activate internal resources of an individual, stimulate appropriate positive behavior, and develop skills: sociability and activity, altruism, and positive attitudes towards themselves and others; they also strengthen the immune system of the individual and improve conflict resolution skills. However, even though a lot of attention is devoted to research on psychological welfare, which is analyzed on many levels around the world, some results are hard to explain and thus cannot be directly relied on when forecasting or modeling the strengthening processes or strategies of subjective welfare in different sociocultural areas and countries.

Psychological welfare and happiness is the object of health psychology studies in many countries. Intensive studies have been carried out in this area since 1997, when it was found that various consequences of stressors (reduced productivity, initiative, etc.), for example, in the USA, determine $100 billion losses per year. The aim of this research is to provide practical recommendations on how to improve life quality, and to create healthy living surroundings, as only in such an environment can high economic and social indices be achieved—i.e., only in such an environment is there a possibility to reach all the possible economic, social, and environmental objectives of sustainable development.

In 2006, the British environmental organization New Economics Foundation (NEF) published the first Happy Planet report, in which the Happy Planet Index was calculated for 178 countries. At the top of this list were countries in which people felt happy as well as having certain objective reasons to feel this way. In the 2006 Happy Planet Index, first place was occupied by Vanuatu island in the Pacific

Ocean, while Great Britain took only the 108th position, and Norway 115th. Among the Baltic states, Lithuania took the lead (149), Latvia took 160th place, and Estonia 173rd. The top of the list was occupied by countries of Central America, while the last ten places were filled with African and Eastern European countries. Germany took 81st place, Japan 95th, and the USA was only 150th.

The authors of the study employed the original evaluation methodology. When evaluating countries according to happiness, usually data about the size of GDP per capita or similar indicators, predicting life expectancy or revealing the environmental pollution level, are used from different countries. In the Happy Planet Index, various social indicators are evaluated; however, the evaluation of their own life by individual people living in different countries is viewed as the most significant criterion:

$$\text{Index} = (\text{life satisfaction} \times \text{life expectancy})/\text{environmental pollution}$$

Among the top ten countries in the 2006 ranking, where residents are the most satisfied with their lives, we also find (in order): Colombia, Costa Rica, Dominican Republic, Panama, Cuba, Honduras, Guatemala, and Salvador. Malta (40) took the highest place among the European countries, while Austria, Iceland, Switzerland, and Italy was at the beginning of the sixth tenth of this list. These results are not that cheerful, but understandable. Intensive work of a large group of scientists revealed a very simple truth: the possibility of unrestricted consumerism and a comfortable life, filled with modern technologies, does not guarantee an individual's well-being and a feeling of happiness.

In addition, another important indicator, the human freedom index, calculated by NEF, also clearly indicated that high rates of GDP per capita do not guarantee the sufficient freedom of individuals. Although there is a tendency that a higher GDP per capita provides a higher index of freedom, there are exceptions; for instance, Estonia took first place according to the index of freedom (85.25) but, according to GDP per capita, it was only in 43rd place in the world. Moreover, Luxembourg had the highest GDP ($69,800) but was only in eleventh place according to the index of freedom (80.09). In Equatorial Guinea, GDP per capita ($16,507) slightly exceeded the rates of Estonia, but this country was only in 150th place based on the index of freedom (26.07). Qatar took eleventh place according to the GDP per capita, and 112th place on the basis of the index of freedom. Furthermore, according to the Happy Planet Index, Vanuatu was only in 207th place among 233 economies in the world based on GDP per capita.

When commenting on the Happy Planet Index, Richard Layard, the director of the Well-being Program at the famous London School of Economics, emphasized that, despite economic growth, happiness in the West has not grown in the last 50 years (Layard 2005). It is important to understand that binge consumption and binge production, corresponding to its demands, cause catastrophic environmental pollution and exhaust nature, while the reckless consumption of life causes a similarly catastrophic pollution of consciousness.

Thus, as shown in the study, people can live long and happy lives without using many natural resources. One of the Happy Planet Index creators, Nic Marks, stresses that the purpose of this index is to show that well-being should not be associated with a high standard of living, and with consumption level. It is obvious that even the countries in the top places in the Happy Planet Index are not perfect. However, the ranking of countries according to the Happy Planet Index shows that all people can easily achieve a long and happy life, if they live by their environmental limits. The tiny country called Vanuatu, located in the middle of the South Pacific Ocean, has a population of only 209,000 people, and its economy is based on tourism and small-scale farming. However, the residents of the country are happy as they can have a long and happy life in an uncontaminated environment and enjoy the pleasures of life, which are not related to binge consumption.

As the economy campaigner Simon Bullock from the organization Friends of the Earth states, the study showed that happiness should not cost the resources of Earth (Bullock 2001).

NEF encourages acceptance of the global Happiness Manifesto, which lists recommendations on how countries and nations could live within their environmental limits and, in this way, would improve their residents' quality of life. Some of the most important recommendations of the Happiness Manifesto are: to eradicate extreme poverty and hunger; to acknowledge individuals' contribution to the common well-being and to appreciate unpaid work; to ensure that the economic development would remain within the environmental limits (Bullock 2001). In its study, NEF warns that, if the global level of consumption corresponded to that of Great Britain, we would need 3.1 times more land to meet this demand.

Another important piece of research, which was prepared by scientists working at the University of Leicester in 2006, also identified the happiest countries in accordance with their own evaluation methodology. Based on the data of this research, the happiest country in the world is Denmark. Meanwhile, Lithuania was among the 30th unhappiest countries, taking the 155th position from 178. However, it was not that far behind its neighboring countries and even managed to overtake some of them. Latvia was in 154th place, Belarus in 170th place, while Russia took 167th place. Only Poland was evaluated much better. It took 99th place, while Estonia was 139th.

According to the researchers, the most unfortunate places to live are African countries: Congo, Zimbabwe, and Burundi. For this study, the social psychologist from the University of Leicester Adrian White and his team used data from various studies carried out by the United Nations, the World Economic Forum, and the World Health Organization, as well as other international organizations.

White states, "We care about more things than whether you are satisfied with your life." The most important question raised by the authors of the study was if people are satisfied with their current situation or environment. According to the research, the key factors influencing happiness are health protection, material welfare, and education. In addition, the researchers created the first World Map of Happiness. In this map, the other happiest countries after Denmark are as follows: Switzerland, Austria, Iceland, and the Bahamas. The United States of America takes 23rd place, Great Britain 41st, Germany 35th, and France is in 62nd. Countries involved in various military or other conflicts, for example, Iraq, are not included in this list. Smaller countries are often considered to be happier, as they have a stronger sense of collectivity; in addition, the aesthetic image of the country has a strong impact on the happiness of its people. The researchers from the University of Leicester were surprised that some Asian countries ended up at the bottom of the list: China took 82nd place, Japan 90th, while India was in 125th position. "It was believed that these countries had a strong collective identity, which by other scientists is also associated with well-being," stated British scientist Adrian White, who admits that his research data are not quite accurate. Regularly using the same methods all around the world would give scientists an opportunity to better understand what factors lead to happiness. The researcher hopes that in the future all countries will be able to perform this type of research at least twice a year.

Meanwhile, in another study (Inglehart and Klingemann 2000), the indices of happiness of Lithuania, Estonia, and Latvia constituted 49, 52, and 55 points, respectively. Ukraine, Belarus, and Russia were among the unhappiest countries, with their happiness indices being less than 32 points; the highest index of happiness was identified in Iceland, which gained 94 points. The Netherlands, Denmark, and Switzerland were not that far behind Iceland. In addition, all Scandinavian countries were near the top of the list. Among the poorest countries, the highest index of happiness was identified in Puerto Rico, while Taiwan, South Korea, Venezuela, and Columbia were close behind. In this study, all post-communist countries belong to the group of unhappiest countries, whereas the rich Western European countries belong to the group of the happiest countries.

Various surveys, such as the *World Value Survey* or *Gallup Survey*, indicate that Lithuania is among the countries with the lowest indices of happiness and the highest suicide rates. At Erasmus University Rotterdam, scientists, led by Ruut Veenhoven, established the Erasmus Happiness Economics Research Organization. Based on the World Database of Happiness, created by this organization, the happiness of Lithuanians is almost 40 percent below the highest possible happiness score (Table 1.1).

For the time being, the biggest scourge in Lithuania's health care system is suicides. During the pre-war times, the number of suicides was very low due to religion because believers viewed suicides as mortal sins; after the restoration of independence, this curve significantly increased. During the biggest recession, Lithuania would lose about 1,600 people every year. Later, this rate decreased to

**TABLE 1.1**
**Data on the Happiness of Population**

| Happiness | Average happiness | Happy life years | Inequality of happiness | Inequality adjusted happiness |
|---|---|---|---|---|
| *Possible ranges* | 0–10 | 0–100 | 0–3.5 | 0–100 |
| *Highest score* | 8.5 Costa Rica | 66.7 Costa Rica | 1.42 Netherlands | 73 Denmark |
| *Lithuania* | 5.5 | 40.0 | 2.24 | 44 |
| *Lowest score* | 2.6 | 12.5 | 3.19 | 16 |

*Source:* created by authors based on http://worlddatabaseofhappiness.eur.nl, 2013.

1,200–1,100 people and in 2012 it amounted to 930 people; however, in 2013, the number increased to 1,100 people. Later, it emerged that in 2013 an additional 100 people took their own lives (Lekecinskaite and Lesinskiene 2017). Based on the data of the World Health Organization (2014), in Lithuania, the number of suicides per 100,000 people in a year constitutes 61.3 men and 10.4 women.

The World Health Organization (WHO), including not only the EU countries but also Norway and Switzerland, concluded that approximately one-third of people had a certain episode of a mental disorder within a year. About 1–2 percent of people were diagnosed with psychosis, whereas about 5.6 percent of men and 1.3 percent of women were discovered to have problems with various addictions. Even though the average life expectancy is increasing, the population is aging; thus, more cases of dementia are diagnosed. At the age of 80, every fifth person is diagnosed with dementia. Mental illnesses are one of the main causes of disability in Europe. The WHO predicts that by 2030 mental illnesses will take first place in the world among chronic diseases leading to disabilities. However, in Lithuania, no one has carried out such epidemiological research. It is known that approximately 5.4 percent of the population (about 164,000 people) were diagnosed with some kind of mental disorder. Moreover, around 55,000 people were diagnosed with alcohol dependence. However, this is only a small number of the people suffering from alcohol consumption. Taking into account the amount of identified alcohol-induced psychoses, it can be assumed that about 150,000 Lithuanians are dependent on alcohol. This is one of the main causes of suicide.

A couple of decades ago, it was possible to propose a hypothesis that such indicators may be scientifically explained on the basis of E. Durkheim's theory on the impact of political changes and the economic crisis on the mental health of society, as well as on G. E. Cornia's insights about the fluctuations in human welfare during the transition periods of the economy. Though the Lithuanian economy has rapidly increased over the past two decades, while the political system has stabilized, in comparison to 1991, suicide rates of the Lithuanian population have remained almost unchanged. In addition, the index of happiness has stayed almost the same. According to the World Database of Happiness created by Erasmus University Rotterdam, the subjective happiness of the Lithuanian population has seen very little change since 2001 (Figure 1.1).

The indicators of happiness of the Lithuanian population, which have remained almost the same over the last 20 years even though the economy has boomed, show that new scientific explanations of happiness as a phenomenon and constructive scientific and political decisions are essential in order to ensure happiness and life satisfaction of the residents of the country. Research on health and positive psychology reveal that the unhappiness of the residents and factors related to it, such as self-harming behavior (irresponsible consumption of alcohol, smoking, and suicides), are detrimental to the economy of the country and social development of the society. What constructive solutions would help to increase the happiness of the population and subjective welfare that reflects it?

Over the last decade, researchers in the field of positive psychology have distinguished factors influencing subjective welfare; in other words, the human qualities which determine the subjective welfare of the person and how long the person will

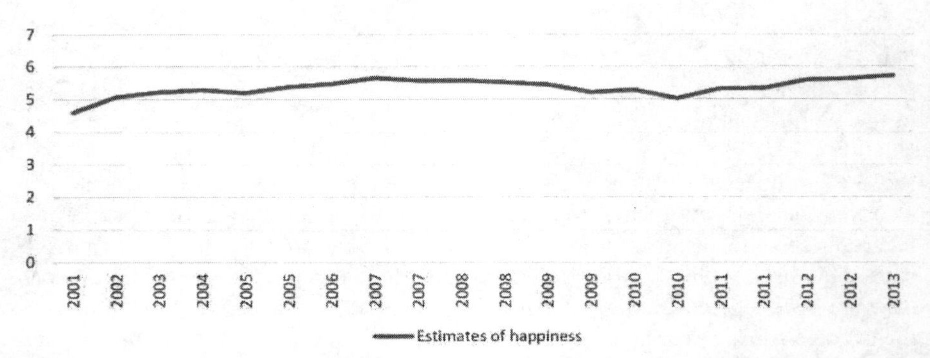


**FIGURE 1.1** Dynamics of subjective happiness evaluation of Lithuanian residents. (Source: created by authors based on: http://worlddatabaseofhappiness.eur.nl.)

be able to take care of his subjective welfare in order to be happy. One of these factors is psychological capital, which is comprised of self-efficiency, hope, optimism, and resistance. Psychological capital determines cognitive, emotional, and behavioral consequences, including the psychological welfare of a human. Research shows that people with high psychological capital also have a high level of psychological welfare and are better prepared to overcome the challenges and stresses of the changing environment, as they are open to new experiences, flexible in dealing with complex and diverse problems, and more emotionally stable when faced with life challenges and difficulties. In addition, many studies have confirmed that psychological capital is not genetically determined, meaning that it is not a constant and innate human characteristic; this means that it may be changed, strengthened, and developed. Thus, changing and developing psychological capital of the people may also change the level of subjective welfare in the country.

Nevertheless, the discussed indices of happiness first reflect personal views of the index compilers about what constitutes a happy person and factors determining happiness. Actually, only the person can tell if he is happy or not. Only a person can evaluate to what extent one or another thing can make him happy or unhappy. Therefore, we can say that material wealth does not bring happiness to a person and he should not seek it in order to be happy. However, the economy cannot speak about happiness; it can only talk about economic choices. Of course, it is easy to see that many people choose to have more and not fewer economic goods; however, in every particular situation, a person chooses what is best for him: material or spiritual values, glimpses of the outer world or the depths of the inner world, as well as humanistic or other values and principles.

Countries' economic growth, industrialization, and capitalism have very little in common with the feeling of happiness inside the human being. For many people, happiness is either unknown or, vice versa, discovered in very simple things. Research shows that owning a cat or a dog helps to improve happiness even more than having a good close personal relationship. There may be two answers to the question of what to do to improve happiness: either you have to wish for what you already have or have something that you wish for. Therefore, it is natural that a rich person who owns a sports car worth millions of dollars will not be happy if he distresses himself that his car is not the best or his neighbor has a more expensive car. Conversely, a person who owns a cheap bicycle will be happy if this type of transport is acceptable to him and if it satisfies his needs.

The market economy can offer only one solution for those who seek happiness: better conditions to have more if you have already chosen a strategy to constantly chase your desires. Nevertheless, this does not mean that the provided opportunities will be used properly. Intelligence, diligence, humanistic values, and an atmosphere of trust in society—these are the things that let us spread the fruits of prosperity in specific countries. Most importantly, the pursuit of economic welfare is not a meaningful strategy to enhance your personal happiness. Usually, after the first wishes are satisfied, other ones arise; then other wishes are thought of and, after that, there are some more of them. If once shoes were seen as a luxury, after that the luxury becomes good shoes, then two pairs of different shoes, and later fashionable shoes, and finally a wide selection of different fashionable shoes for every occasion. But it is obvious that a barefoot shepherd may be happier than a lady who has a huge shoe collection or who has inherited enormous wealth.

It has to be noted that, according to the latest data, in the lists of the happiest nations, Lithuanians are very low—they take 155th place in the list from the University of Leicester and 149th place in the report of the New Economics Foundation. Meanwhile, in the reports of the United Nations Development Programme on Human Social Development, which calculates the index of human development in 177 countries worldwide, Lithuania is in the first quarter of the most advanced countries. Thus, for more than a decade, Lithuania has established itself among countries that have a high index of social development. Of course, this indicator, as with others, does not show or explain everything; it is deduced from the purchasing power, the number of educated people, and the average life expectancy. However, this is an official and representative indicator which is respected, not like other social studies, and is based not on speculative opinions, but on completely real and verifiable data. It includes not only tangible but also intangible data and evaluates welfare in a broader sense. For example, an average life expectancy says a lot not only about the health care and living and working conditions, but also about the will and the desire to live: everybody knows that, without the motivation to live, people will die faster.

The authors of this research made their own conclusions that pessimism can be caused not only by living conditions but also by the biological properties of the brain. Lithuanians usually state that the reason why they are unhappy is that they do not earn enough money and can afford only little in comparison with other EU countries. But the fact is that, "according to happiness," we are almost at the very bottom of the list, whereas in the global context our material status is more than satisfactory. Therefore, why, with life getting better, do we not feel any happier? Clearly, by asking this question, many important factors in a person's life are being ignored.

For a number of years, Norway and Iceland have taken first place in the Social Progress Index, while Sweden has taken second place, as their indices of happiness are excellent. Lithuanians are somewhat ahead of Russians, Byelorussians, and Ukrainians, who are ahead of the central African countries, which are at the bottom of the list.

We all understand why the Danes, Swiss, Icelanders, and Austrians are happy, but Brunei and Bhutan step in between these nations; Brunei is a Muslim country in South Asia ruled by a sultan. It is 12 times smaller than Lithuania, its main exports are oil and gas, and it has a very dense population that lives in poorer conditions than the population of Zimbabwe, which is at the very bottom of the list. This means that some nations are able to adjust to difficult conditions, and for others, the same conditions are a disaster. Happy Bhutan is an economically underdeveloped Buddhist country. Neither welfare, nor good climate, nor religion are the absolute conditions to be happy. Even if these three conditions are combined, they do not guarantee happiness for some countries.

Thus, whether people are happy or not should be a natural condition arising during the evaluation of the essence of their lives. Usually, a person does not go too deep into this topic and pores about it only when he is asked to; however, his face, behavior and pace tell everything else about him. If we would smile in front of a mirror every single morning, as advised by psychologists, the streets would definitely look nicer, even though this would be only a happiness exercise. But what should be done in order to make all Lithuanians happy? How can we adjust the broken mechanisms of happiness in the hearts of Lithuanians? Neither the sunsets, nor the sunrises, nor the chirps of birds influence a modern person. It seems that Lithuanians, as a nation, are suffering from one of the biggest syndromes in the world, which is the syndrome of being a victim. One of the most important discussion topics is the freedom of an individual in this changing world, as freedom is directly associated with happiness. From a position of traditionally understood progress, the person should be freer with every step.

Unfortunately, the experience of the last couple of decades allows us to quite skeptically evaluate the optimistic vision of an ever-improving person. Suppose one of the oldest democracies, Great Britain, has the highest number of cameras watching people in public and private places. In the USA, a bastion of Western democracy and individual freedoms, citizens' right to privacy will be severely limited by the initiative of the federal government. Thus, with the modernization of society, greater efforts are required to fight crime, and the number of people suffering from mental disorders is increasing, as are the number of suicides and global environmental problems, such as climate change. In addition, the most advantageous preconditions for the rise of international terrorism have emerged. Hence, the analysis of these problems would allow us to justify further research on happiness and to find the best solutions to these problems.

However, economic growth and the growth of the population's economic welfare cannot guarantee happiness, social welfare, and environmental conservation for the present and further generations. Thus, the harmony between economic, social, and environmental objectives is a must in order to realize policy measures and to seek sustainable development of the country. As the economy develops, it has a negative impact on the environment, but this competitive economy is primarily based on new, resource-efficient technologies and the realization of it leads to a significant reduction in the negative impact on the environment. Only healthy, educated, and happy people can ensure economic growth and create preconditions for the competitive economy. Hence, the investment in human resources, secured and trust-based social, working and domestic environment at the same time are key factors for sustainable development.

It can be stated that sustainable development is primarily determined by the economic relationship with the happiness of the residents; consequently, the main task of the policies is to create not financial capital, but the main asset of the country: an equal, healthy, and vibrant population in which people are happy with their activities at work and during their leisure time. However, there is a problem of incompetence that exists at all stages of the development of society. Leadership and developed principles of leadership are required to tackle this problem. Without effective governance or a strong institutional dimension, it is impossible to create an equal, healthy, and vibrant society as well as to guarantee its sustainable development.

## 1.3 SUSTAINABILITY MEASUREMENT

The methods of sustainability measurement may be placed into four main groups (Harger and Meyer 1996): indicators and indices, and sustainability evaluation means at the product level (production methods), at the project level, and at the country level. Moreover, all the measurement methods may be grouped in accordance with their dimensions of sustainable development (environmental, social, economic, integrated, and covering all dimensions of sustainable development). Table 1.2 is grouped by the methods and tools which are applicable to sustainability measurement.

The first group of measures used to assess sustainability consists of indicators. Indicators are a simple tool which enable assessment of the economic, social, and environmental purposes of a country's development. If the environmental, social, and economic indicators are integrated into one indicator, they constitute the index. Indicators have to distinguish themselves with such important features as simplicity, wide coverage, and quantification possibility that allows determining tendencies. The tendency evaluation allows for the performance of short-term forecasts (Table 1.2). All the indicators of the categories may be grouped into non-integrated and integrated indicators (indices). An individual class is constituted of the regional flow indicators.

The example of a non-integrated indicator is Environmental Pressure Indicators (EPI), prepared by the European Union Statistics Agency Eurostat (European Commission and Eurostat 1999, 2001). Eurostat collects these indicators for the EU Member States and regions, in cooperation with the EU Member States' statistics departments. EPI forms 60 indicators, six indicators for each sphere of the environmental policy, established in accordance with the Fifth Environmental Action Program (ten policy spheres). These six indicators may be aggregated into indices in each policy area, which in turn consist of ten indices of the environmental pressure. The entirety of these indicators, which consist of, for example, damage to forests, tourism intensity, contaminated soil, contaminated water, etc., is appointed for the evaluation of the EU Member States' environmental sustainability. The indicators allow countries to be compared and their tendencies to be evaluated in the main policy spheres.

**TABLE 1.2**
**The Methods and Tools of Sustainability Measurement**

| | Indicators/indices | The evaluation of products and technologies | The evaluation of projects | Sectoral, country-level evaluation |
|---|---|---|---|---|
| The environmental dimension | The indicators of environmental pressure<br>Ecological footprint<br>Environmental space | Life-cycle evaluation;<br>The material input per unit of service;<br>The material flow analysis;<br>The energy flow analysis;<br>The exergy analysis;<br>The emergy analysis | The environmental impact evaluation;<br>The ecological risk analysis | Environmental extended inter-branch balance;<br>Inter-branch energy balance;<br>Strategic environmental impact evaluation;<br>Regional emergy analysis;<br>Regional exergy analysis |
| The economic dimension | The total extent of the national production | Life-cycle expenditure | Entire life-cycle expenditure accounting | Economical material flow analysis;<br>Economical energy flow analysis;<br>Economic inter-branch balance |
| The social dimension | Social indicators | | Evaluation of the social impact | Social inter-branch balance |
| An integrated method | Human Development Index;<br>GINI coefficient;<br>The Environmental Sustainability Index;<br>Welfare index;<br>Sustainable national income;<br>Genuine Progress indicator;<br>Happy Planet Index;<br>The sustainability barometer;<br>Fair savings indicator | | Cost–benefit analysis;<br>Risk analysis | Multi-criteria analysis;<br>Vulnerability analysis |
| Sustainable development | UN sustainable development indicators;<br>The sustainability indicators of individual economic countries (energetics, transport, agriculture, and tourism) | | | Conceptual modeling;<br>Systems dynamics;<br>Sustainability impact evaluation;<br>Integrated sustainability evaluation |

*Source:* created by authors.

The other example of the non-integrated indicators is the set of 58 sustainability indicators, used by the UN Commission on Sustainable Development (Lammers and Gilbert 1999). These indicators cover not only economic, environmental, and social dimensions, but also the institutional dimension. Such a system of indicators is not integrated. Examples may be the water quality level, national literacy rate, pace of population growth, GDP per capita, and the number of ratified international agreements, etc. From 1994, on the basis of these indicators, countries prepare reports to the UN and account for the results of the sustainable development principles realization (UNCSD 2001). Moreover, these indicators are used to prepare the strategies of countries' sustainable development.

Various systems of the sustainable development indicators are created for the individual economic branches: energetics, transport, industry, tourism, agriculture, etc. For example, the Sustainable Energy Development Indicators cover social, economic, and environmental energy sector dimensions. The indicator system was formed by the efforts of International Atomic Energy Agency Eurostat and the United Nations Commission on Sustainable Development. These indicators are not integrated and are used to evaluate the sustainability of a country's energy sector. Using these indicators, the evaluation of sustainable energy development tendency is allowed, and respective actions may be taken to change or promote these tendencies. This system also allows countries to be compared in accordance with an individual indicator and development tendencies of their energy sector to be evaluated.

The material and energy flow analysis allows the structure of the resource flow to be analyzed and the inefficiency manifestations in a system to be found. This class of indicators may be used for historical flow and the emission tendency analysis as well as in decision-making. Material flow analysis (MFA) analyzes physical society metabolism, in order to support dematerialization processes and to reduce the negative impact on the environment, which is related to the usage of wasteful resources (United Nations 2002). MFA studies have been performed in many countries; moreover, the number of performed regional MFA studies has significantly increased in recent decades. This class of indicators is the class of non-integrated indicators; they only cover physical flow, thus only the environmental dimension. The entire economic MFA, estimated by Eurostat, is a standard tool of MFA, applied to compare the EU Member States. World Resources Institute MFA studies for the developed world countries were the initial tool to standardize MFA in the EU (Kleijn 2001). Eurostat has prepared landmarks for the MFA economy evaluation. Eurostat landmarks on the material flow indicators are divided into three categories: input, outflow, and consumption (Anderberg, Prieler, Olendrzynski, and de Bruyn 2000). Each category covers different levels; depending on the level, it covers local, foreign, or concealed flows (Adriaanse et al. 1997). Concealed flows are those which are not included in the economic system—for example, soil erosion etc.

Material input indicators show the material inflows into the economy via local production and consumption. Material outflow indicators measure outfalls of all materials back to the environment or evaluate all pollutants which are discarded into the environment during the time of manufacture or consumption process. Material consumption indicators measure all materials consumed in the economy.

Substance flow analysis (SFA) covers certain chemicals of regional flow and negative impact in relation therewith to the environment or environment loss. The aim of the SFA is to reduce the environmental pressures which are done by a certain material consumption. SFA analysis is done at a regional or a country level, in order to determine problematical areas and to select necessary measures in problem-solving. It is useful for planning environmental policy and decision-making.

Energy flow analysis covers energy flows in the economy. It is based on the first law of thermodynamics or energy persistence, which states that energy quantity has a constant size and energy can be neither created nor destroyed, and can only move from one shape to another (Hovelius 1997). Country or region energy analysis is carried out using input-output energy flow analysis, which is based on Wassily Leontief's economic input-output matrix, which analyzes flow between the individual industries in the economy. In the case of energy analysis, marketing flows are replaced by energy flows between industries or sectors.

In addition, energy analysis may be performed using exergy and emergy analysis. This analysis is more advanced because during it both the energy quality and its quantity are assessed (Herendeen 2004). System exergy is the maximum amount of mechanical work quantity which may be obtained in that system. Exergy analysis shows the efficiency of resource use and can determine where losses form and where technological improvements may be carried out to increase the efficiency of energy usage. Regional exergy analyses were carried out in Sweden, Japan, and the USA (Wall 1990, 1997; Ayres, Ayres, and Warr 2003). Their results allowed the methodology of regional exergy analysis to be prepared, which is based on the expression of all resources and goods on a single unit of measurement, emjoules, which is the amount of solar energy they need to produce them.

Currently, around the world, many efforts are put into integrating sustainable development indicators, to create one single index which reflects the achievements of sustainable development. The first efforts were directed at supplementing or creating the indices of a new national reporting system such as gross domestic product (GDP), gross national product (GNP), and net national income (NNI) to assess the common welfare, achieved by the country. Traditional indicators (GDP, GNP, and NNI) sent the wrong signals about the reached welfare level in the country, without underestimating the achievements of sustainable development, such as inequality of income distribution, public security, resource over-wasting or not evaluated external effects (positive or negative external costs). Because of GDP limitations, the need to assess the environmental dimension, and a willingness to determine an adequate quality of life indicator, many GDP modifications were proposed. All of them were designated to assess the effectiveness of the country's sustainable development.

To do a complex evaluation of sustainable development, it is important to connect four dimensions (economic, social, environmental, and institutional); however, in scientific literature, it is observed that the indicators of sustainable development are usually connected only into two groups. Currently, around the world, more than 500 sustainable development indicators are counted which have been developed by governmental and intergovernmental organizations for their own needs: about 70 of them are global, over 100 are national, more than 70 are regional, and about 300 are local.

As stated in the UN's publication *Sustainable Consumption and Production*, published in 2007, sustainable development indicators perform many functions. They may improve decisions and make much more effective measures, simplifying, clarifying, and making summaries, which are accessible and understandable to politicians (Ciegis, Ramanauskiene, and Martinkus 2009). The aim of the indicator is to show how well the system works. If a problem occurs, the indicator may help to determine in which direction the problem needs to be solved. The choice of specific sustainability indicators (or indices) should be guided by the following principles of sustainable development: social justice; local self-government; public involvement, democracy; sustainable balance between the usage of local and imported resources; the usage of local economics potential; environmental protection; cultural heritage conservation; the new environment quality protection and restoration, functionality and attractiveness magnification of operated space and buildings (Ciegis, Ramanauskiene, and Martinkus 2009).

Indicators are used in different cases, in many spheres of life and activities. They transform available knowledge into a user-friendly form, thereby facilitating the decision-making process, depending on the information required in the sustainable development process. In order to manage sustainability, society has to formulate clear and measurable sustainability objectives, which must be constantly reconsidered and revised. The degree to which these goals are implemented might be measured using sustainable development indicators—definable and measurable parameters, the value and the direction of change which show a specific indicator of ecological, economic, and social stability development.

Having in mind that the main aim of sustainable development is related to the compatibility of economic, social, and environmental sphere goals, compatibility is mostly orientated into the selection of appropriate indicators and their evaluation. The shortage (uncertainty) of sustainability conception is exposed here: when interpreting sustainable development, attention is directed to the indicators, forgetting other important aspects. For example, indicators transfer only certain information but do not specify how and in what ways the sustainability may be reached.

The sustainable development evaluation may be performed combining quantitative indicators with qualitative information because, in this case, the weakest spots and measures will be specified which will improve the current situation (Krajnc and Glavic 2005). Although there are many various sustainable development indicators, they do not cover and describe everything properly, so the complex sustainable development evaluation needs to be carried out to distinguish the most essential indicators. In the scientific literature, in terms of sustainable development indicators, the transition from a necessary collection of information to the construction of indices is noticeable. The first stage is oriented towards the collection and processing of information because the plain and non-systematized information is not appropriate in certain decision-making. After collecting the necessary data, indicators are counted and indices are later aggregated.

Therefore, the data are first distinguished, and then indicators, sets of indicators, and finally indices are formed. This creates the information pyramid with an aggregate index on top which covers all sustainable development dimensions. However, at first, an introduction with the conception of indicators should be made, which allows the action plan to be concluded to properly organize a work process, the final result of which is the expression of various indicators in a single index.

Sustainable development indicators may be described as the combination of the quantitative and qualitative indicators which formalizes sustainable development itself. These indicators are measures which provide beneficial information about the physical, social, or economic condition. An indicator is a size that differs from the normally measured values (mathematical values), as it is typical to meet certain goals (to give results) beyond what was directly measured. Therefore, the final indices should be composed only of a few numbers; otherwise, decision-makers and society will have difficulty interpreting and using them (Ciegis, Ramanauskiene, and Martinkus 2009).

The classification of indicators into subjective/objective and quantitative/qualitative groups usually shows the expression of quantitative indicators by numbers, while qualitative indicators are those that are difficult to express in numbers. Objective indicators are those to the result of which a person himself/herself will not have any influence, while subjective indicators are already dependent on a person's attitude, his/her mood, and understanding. Therefore, in accordance with these aspects, four different types of sustainable development indicators (quantitative, qualitative, objective, and subjective) may be formed.

The main advantages of the complex index are distinguished: they show how different indicators are interrelated; give insights on the dimension of social economic environmental indicators; and allow general sustainable development projections (long-term trends) to be performed (Juknys 2008; Ciegis, Ramanauskiene, and Martinkus 2009; Staniskis, Arbaciauskas, Stasiskiene, and Varzinskas 2007; Staniskis and Stasiskiene 2008).

The creation of the integrated indicator should not cause difficulties while doing comparisons—for example, one common index has to provide simplified quantitative information about sustainable development. Sustainable development indicators may be expressed not only in numbers, but there may be certain signs or labels which will increase the understanding about the environment (what is happening there), and by analyzing them, the necessary decisions could be accepted, or the action plans could be made. Although the complex indicators that are being developed provide useful information about sustainable development tendencies and the sustainable development key issues, they also have some drawbacks (Ciegis, Ramanauskiene, and Startiene 2009).

The following major drawbacks of complex indices may be specified:

- They do not match the simplicity criterion. This means that the assumptions of methodical calculation are not fully understandable; therefore, society's participation possibilities to make decisions on the basis of such indicators are very limited, and this reduces the transparency of the made decisions.
- It is also noticeable that, in order to facilitate information analysis and data presentation, a graphic way to depict data is required.

However, individual and aggregated indicators do not foresee exchanges among the three most important sizes: efficiency, justice, and sustainability. A list of finite indicators which does not change could not be prepared because various changes both in particular countries and around the world are constantly happening (Bass and Dalal-Clayton 2002). Nevertheless, the system of indicators itself is necessary in order to implement the formulated sustainable development tasks and to perform their control.

In summary, the complex indicators mostly cover only two or three dimensions; therefore, complex sustainable development indicators and dimensions are distinguished (Figure 1.2) (Suri and Chapman 1998; Schneider and Kay 1994; Grossman and Krueger 1991; Bruyn 2000).

Economic and social dimensions are reflected by two most commonly used indicators: the Human Development Index and GINI coefficient.

Since 1975, the United Nations Development Programme (UNDP) has published annual global reports on human development (United Nations Development Programme 2011). The report counts the Human Development Index (HDI) of 175 countries, according to which, in 2014, Norway was ranked first out of 187 countries, with an HDI of 0.943 (the average life expectancy was 81.1 years, and GDP per capita was 47,500,000 of purchasing power standard [PPS in 2009, $]). Meanwhile, Lithuania took 40th place, with an HDI of 0.810 (the average life expectancy was 72.2 years, and the GDP contained about 16,200,000 PPS). It takes two years to count HDI because it is difficult to evaluate this year's or last year's results achieved by countries.

The Human Development Index is an important measure of sustainable development and, in particular, of human development. It allows assessment of the social

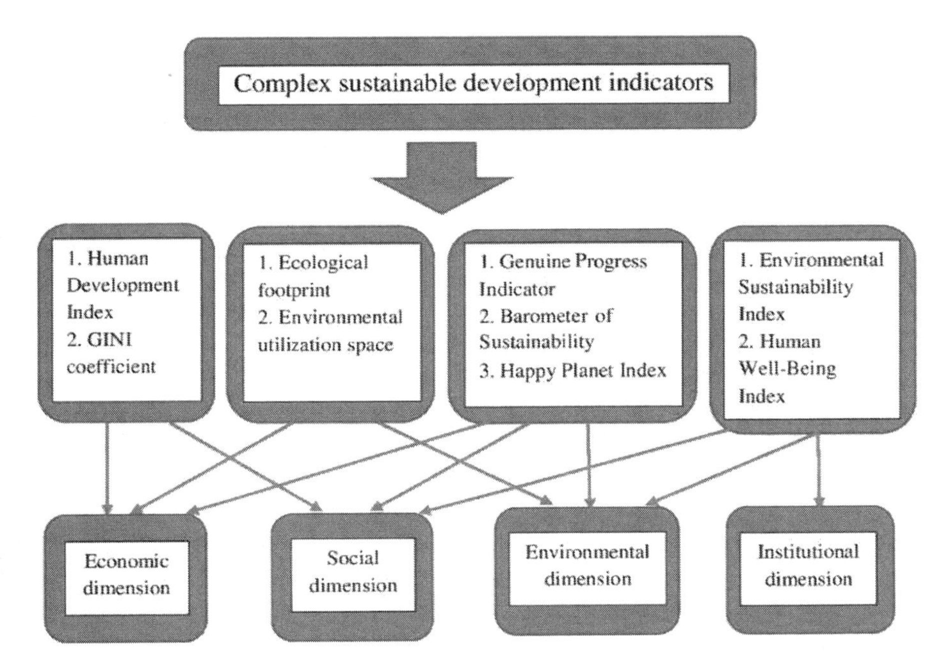


**FIGURE 1.2** Complex sustainable development indicators. (Source: created by authors.)

living conditions in a country. This index deals with the social standard of living in a particular country. The index is used to determine if a state is developed and allows the influence of the country's economic policy on the quality of life to be measured. In 1990, the index was created by Nobel prize winner Amartya Sen and Pakistani Mahbub ul Haq, with the help of American Gustav Ranis and Meghnad Desai, from the UK. Since its creation, the index has been used in the United Nations Development Programme, presenting an annual report on human development in the countries of the world. Interestingly, the index creator himself, Amartya Sen, criticized the index for its limitations; however, the Human Development Index is more useful than the previously used GDP per capita. This index laid the foundations for other, more detailed research which is related to human development, presented in the Human Development Report.

HDI measures the country's achievements, taking into account the three main human development components:

- long and healthy life, which testifies the average future life expectancy;
- knowledge, which testifies the adult literacy rate (two-th irds of the component value) and the total coefficient of those who are trying to gain primary, secondary, and higher education (one-third of the component value);
- a good standard of living, which testifies the GDP per capita.

Before counting HDI, each mentioned component indicator must be determined, and then the total HDI is calculable, which is a simple average of three component indicators. The first UNDP Report on human development was introduced in 1990, with the sole purpose that, in the course of the economic debate, formulating policy and doing the propagated job, people would again appear at the center of the development process. Each report on human development is intended for a very relevant issue in the development debate and provides an innovative situation analysis and a number of important policy recommendations (United Nations Development Programme 2010).

It is important to emphasize that HDI evaluates not only the GDP, the main measure of integrated economic dimension, but also social human choice possibilities, which cover choice possibilities to live long and healthy lives, acquire knowledge, get resources, participate in a society's life, live in a clean environment, etc. However, as specified in the Human Development Report, this indicator evaluates only the basic human choices, as not all aspects of development are quantifiable.

Another indicator which shows human living standard by income and expenditure distribution, their structural changes, and transferred power is the GINI coefficient (in 1912, it was proposed by Italian Statistician C. Gini). This coefficient depicts the share of accrued income received by households, sorted in order from receiving the lowest to the highest incomes (Dixon, Wiener, Mitchell-Olds, and Woodley, 1988). In the world, it was started to calculate income inequality not that long ago; the first calculations and analytical works appeared in 1980 because it was impossible to obtain data from the regions until then.

Having in mind that the complex indices often include indicators, which may repeat in one index and another, by performing the indicator analysis and determining the indicator dependencies, individual indicators are included, which are mostly calculated to the complex indices. Thus, the first group of complex indices previously distinguished is usually the following: GDP, the level of education, average life expectancy, and personal consumption expenditures per household member per month.

Another group of indicators covers economic and environmental dimensions. This group includes the following indices: ecological footprint and environmental utilization space. The ecological footprint is the most commonly used complex indicator, which allows experience of the usage and waste assimilation needs of a certain community of people or economic resources when taking into account the corresponding productive land area. The ecological footprint calculating result is the land area per capita needed for annual consumption of goods and services. The ecological footprint calculation consists of several steps. In the beginning, the average annual consumption of food, living space, transport, consumer goods, and services per person is calculated. Then, the land area required to produce each of the consumables is calculated, and its environmental impact is assessed in accordance with the required land area. Then, when summing those land areas, the land area is obtained, necessary to meet the annual needs of one resident. It has been calculated for many countries and regions and is intended to assess the sustainability of a country, but it may be applied to a city or region.

In terms of the results of ecological footprint studies (Ciegis 2004; Ciegis and Zeleniutė 2008), which were described in the European Commission document on a *Thematic Strategy for the Sustainable Use of Natural Resources*, published in 2006, it should be emphasized that in 2003 the average ecological footprint was 2.2 ha in the world, although in order not to exceed Earth's biological ability to renew, it should be no larger than 1.8 ha. This means that people around the world consume 25 percent more than the Earth produces in a year, or in other words, it would take a year and three months to renew the resources which are consumed in one year. Research shows that people of the richest countries consume an average of 3.5 times more than the planet's resources permits, and 3.9 ha of land is needed to satisfy one Lithuanian's habits, which is 2.17 times more than the average rate. Thus, if all the countries of the world were divided into two groups—economically developed and developing—economically developed countries would satisfy their own needs by using natural resources (water resources, forests, minerals, etc.) and energy resources (oil, coal, natural gas, etc.) of the developing countries.

Another indicator is environmental utilization space, which covers three main parts: the potential of environment absorbent pollutants; the duration of natural resources exploitation; and the ecological capacity of the local or global environment. Environmental space is defined as the content of energy, fresh water, agricultural resources, non-renewable natural resources, and wood resources, which is consumed and allows assessment of whether such usage of the resources is permissible from the point of view of the sustainable development. Therefore, from a second group of the sustainable development indicators to the sustainable development list, first indicators, which reflect the main aspects: water, earth, air (pollutants emission), renewable, and non-renewable energy sources should be included, because the extent of their use would help to assess how the country is developing towards sustainability.

The third group of indicators covers indicators which include three sustainable development dimensions: economic, social, and environmental. These indicators are Genuine Progress Indicator, Barometer of Sustainability, and Happy Planet Index.

Genuine Progress Indicator, which covers three sustainability dimensions, may be described as the index of welfare and sustainable development, which provides a way of measuring the practical sustainable development progress achieved in a country. This index is based on the fundamental understanding that social, economic, and environmental dimensions are inseparably related. One of the aims of the index appearance is to show the economic growth possibilities, combining them with non-renewable usage of resources. The basis of methodology creation of the Genuine Progress Index is personal consumption expenditures, adjusted by the income inequality; this index also includes more than 20 other indicators reflecting economic life aspects.

The Barometer of Sustainability was invented by several professionals (A. Prescott-Allen made a special contribution to the methodology) from the International Union for Conservation of Nature and from the International Development Centre. This is a sustainability evaluation methodology which reflects the environmental condition (the usage of resources and its extent) and society's welfare.

Another complex sustainability indicator is the Happy Planet Index, which includes life satisfaction, multiplied by life expectancy and divided by ecological footprint. It was first calculated in 2006. In 2009, the New Economics Foundation scientists conducted a study which showed that the population may live long and happy years of their life without using large amounts of natural resources. Thus, happiness does not depend on the amount of used natural resources. During the time of the research, 143 countries around the world were analyzed. The happiness index is measured from 0 to 100 points; the higher the value, the "happier" the country. In the 2009 Happy Planet Index, Costa Rica took first place, and Lithuania took 86th place (HPI = 40.9 points). The new HPI 2.0 index was created, which showed not only the index of happiness (as shown in the previous Happy Planet Indices), but also the missing distance to reach this happiness. However, happiness is very subjective (only the human himself/herself may tell what makes him/her happy); therefore, this indicator is rarely used.

The fourth group of indicators covers indicators which reflect social, environmental, and institutional dimensions. It is said that the institutional environment is also important when assessing sustainable development. This group includes the following indices: Environmental Sustainability Index and Human Well-Being Index.

In 2002, the Environmental Sustainability Index was presented by the World Economic Forum. It is calculated for 142 countries and includes 21 indicators, which have 76 other variables (indicators). The Environmental Sustainability Index shows the overall progress in the sphere of sustainable development (World Economic Forum 2000). Although the index focuses mostly on the environmental dimension, it also includes the social and institutional dimensions, because such indicators as public sector activities, international cooperation, human welfare and other indicators are included. The key Environmental Sustainability Index disadvantages are the following:

- This index does not include relations between separate sustainable development dimensions.
- The index gives preference to developed countries, because new technologies and innovations are included that are not available to developing countries.
- The importance weights to all indicators are provided equally, regardless of their importance to the environment.
- When calculating the Environmental Sustainability Index, economic indicators are not included completely (such as GDP per capita, employment, food consumption, etc.).

One more indicator which is included in the fourth group is the Human Well-Being Index. It is one of the most commonly used complex sustainable development indices, and covers the same dimensions as the Environmental Sustainability Index (social, environmental, and institutional dimensions). This index consists of 36 indicators of health, population, education, communication, freedom, peace, criminality, and equality, which cover social and institutional dimensions and form the Human Well-Being Index; it also consists of 51 indicators of earth, biodiversity, water and air quality, the usage of energy and resources, which cover the environmental dimension and form the Well-Being Index. These two indices form a common Well-Being Index.

Many of these mentioned complex sustainable development indices have advantages (e.g., they represent a certain sustainable development dimension and the current situation); however, they also have disadvantages because they are subjective (e.g., Happiness Index, Well-Being Index), they include many indicators, which can duplicate one another, and they are determined by the same weighting coefficients to the indicators "included" in the indices; therefore, an objective evaluation is required. In this case, the individual indicators, their analysis, and integrated Sustainable Development Index may provide the objectivity.

The abundance of the sustainable development indicators themselves is determined by their necessity for decision-making and management, research, and analysis. Moreover, the choice of indicators means that each indicator has both its strengths and weaknesses and that it is impossible to find one indicator to suit all cases (Ciegis and Ramanauskiene 2011).

In decision-making at the global, national, or regional level, methods for policy measure analysis are applied, and methods of local evaluation are applied at the project level.

Integrated evaluation methods in the context of sustainability evaluation are mainly ex-ante methods. They are usually applied in the form of possible scenario analysis. Many of these integrated evaluation tools are based on system analysis and cover the aspects of nature and society. The integrated evaluation consists of many different tools. Integrated evaluation of environmental problems is carried out using such tools as multi-criteria analysis, risk analysis, vulnerability analysis, and expenditure/benefit analysis.

Conceptual modeling analyzes qualitative (causative) relations and applies flow charts, flow maps, etc. Conceptual modeling may be applied in order to visualize and determine changes in a system, which operate sustainability positively or adapt powerful computer models to conceptualize these relations. System dynamics is a model of complex systems representation created with the help of computer models. They allow experiments to be performed and investigation and observation of the operation of these models over a long period, as well as analysis of various possible scenarios, implementing the different policy measures. Examples of applied models of sustainability evaluation are the general and partial balance models: the models of GEMINI, RAINS, TIMES, BALANCE, IMAGE, etc., to observe and analyze the dynamics of social, biodiversity, and climate systems.

The multi-criteria analysis is applied in order to perform the evaluation of a project or policy measure impact in accordance with a contradictory criterion. Multi-criteria analysis sets the aims and objectives and seeks to weigh them and determine the optimal policy measure in accordance with the aims and objectives. This method allows evaluation of both quantitative and qualitative data. The methodology may be used successfully to form the energetics and environmental policy.

Vulnerability analysis evaluates the human–nature system vulnerability in order to determine how sensitive the system is to the changes and how it is able to cope with the changes. If the analysis shows that the human or environmental system is vulnerable, then the risk analysis is performed. Vulnerability analysis is mostly carried out for the public, ecosystems, energetics, and other infrastructures when investigating the effects of climate change.

Having analyzed the sustainability evaluation methods in accordance with their ability to integrate nature and society, as well as to include long-term and different area levels, it may be said that a few of them integrate nature and society or all three dimensions of sustainability. As seen from the grouping methods in Table 1.2, only integrated methods and sustainable development methods cover all dimensions of sustainable development. Many methods cover only environmental dimensions, particularly at the product level. Only the evaluation of the life-cycle expenditure covers the economic and environmental expenditure of the product.

The aim to expand the analysis is to integrate or connect the methods of sustainability evaluation. For example, the life-cycle evaluation (the method of environmental impact evaluation) was connected with the life-cycle expenditure evaluation (economic method) and social life-cycle evaluation. Although many methods cover the national level, they may also be adapted at lower levels. The methods of product level sustainability evaluation are not locally related; however, the efforts are made to improve these tools, in order to bind the influence in a particular area. Forecasting methods are beneficial in order to determine the impact on sustainability in the long-term perspective. Sustainability evaluation methods should be more standardized and give more transparent results because the latter profusion and diversity bring many uncertainties, evaluating political measures, projects or product sustainability, and choosing between alternative solutions.

As seen in the review of sustainability measurement methods, the main focus, while measuring sustainability, is given to the sustainability measurement of the country, region and organization. In the next section, we will discuss the conception of organizational sustainability and the main organizational sustainability criteria.

## 1.4 ORGANIZATIONAL SUSTAINABILITY

In carrying out their daily activities, each company, organization, or individual is faced with the surrounding environment over which all existing organizations have a higher or lower influence. Therefore, residents, government, and international organizations are forced to supervise that the companies implement effective actions which would have the least negative impact on the environment. Such concern for nature preservation has led businesses to change their behavior in order to prove to the public that business can also be environmentally friendly and socially responsible. This kind of business may contribute to the implementation of the country's objectives of sustainable development and ensure sustainable development at the macro level.

In recent years, organizational sustainability has become one of the most popular and most ambitious modern management concepts. In order to make the organization sustainable, its sustainable development ideas should correspond with the organizational culture and should involve not only the employees and customers, but also the suppliers, partners, and investors as well as drawing the attention of the economic and social development, both in natural resources and environmental preservation.

Economic, social, and environmental sustainability of organizations, including business enterprises, is distinguished. Economic organizational sustainability is understood as a stable, profitable activity of the company and its stable development. Environmental sustainability focuses mostly on stability preservation of the biological and physical systems in order to implement the most important activities of the company. When caring for the environment, the effective use of raw materials and energy, resource saving, and waste recycling are some of the most important environmental sustainability aspects of the business (Simanskienė and Petrulis 2014; Bagdoniene, Galbuogiene, and Paulaviciene 2009). It is important to distinguish the social dimension of sustainability which reflects the connection between the development of the company and the prevailing social norms, and which aims to maintain the stability of social systems, the preservation of cultural diversity, the conflict reduction, and the equality between people and separate generations. The main principle of sustainability is an effective fulfilment of companies' basic needs by protecting the environment and ensuring the social welfare of workers (Atkociuniene and Radiunaite 2011). However, in order to ensure the company's sustainable success, the economic, social, environmental, and organizational values of the company should be integrated into a cohesive whole.

Benefits of sustainability are reflected in all interested parties of the organization: the consumers, who may purchase a product or service which has been made with the consideration of environmental impact reduction, the organization workers, for whom the conducive work conditions are created, and others. The benefits of sustainability for organizations can be distinguished into long-term trust and approval of the society, positive image of the organization, and growth in demand for products or services which ensures the organization's further development and opportunities to work profitably in the future. Another aspect which separates a sustainable organization from others is the willingness to share its created goods in order to ensure the organization's uniqueness, which in return helps to maintain a favorable attitude of the consumers, customers, and government or public institutions. However, organizational sustainability is not a simple phenomenon. It includes eight scientific disciplines: economy, ecology, sociology, business ethics, social contract theory, human rights theory, management, and business law. Economy, ecology, and sociology encompass the concept of sustainable development and take into account the public opinion in achieving the organizations' development goals. Business ethics, social contract theory, and human rights theory encompass the organizations' social responsibility, which includes the organizations' operating philosophy and the aspects related to the organizations' economic, social, and environmental responsibility (Norman 2013). Management is based on stakeholder theory, which is grounded on the organization management by taking into account the needs of stakeholders. Business law is based on organizations' accountability theory, which incorporates business arguments related to the organizations' aim to provide public activity reports.

Currently, the notion of the circular economy has become one of the newest concepts dealing with the issue of environmental sustainability (Murphy and Rosenfield 2016). On the basis of European Commission documents, several ways to achieve the efficient use of resources manifest in the longevity, efficiency, the use of substitutes, eco-design, industrial symbiosis, and leasing or ordinary lease. In order to achieve resource efficiency, all the necessary changes are based on technical, social, and organizational innovations throughout the value chain where production and consumption are combined (Lacy et al. 2014). To accomplish these transformations, the European Commission identifies the following main components: (1) skills and knowledge, including entrepreneurship, capacity building, and diversity; (2) organizational innovations, which include the integrated decisions and systems, logistics, business models, and supplementary political measures; (3) social innovations, which involve new production and consumption patterns, citizens' involvement, product service models, and design services; (4) technological innovations, which encompass materials and processes design, product design, and resource management (e.g., waste, water, energy, and raw materials); (5) financial measures; (6) dissemination and internationalization; (7) the majority of stakeholders. In an attempt to support the investments into the circular economy and to reduce the companies' objections to the emergence of a new economic direction, the European Commission declared that it would demonstrate all possible opportunities in the transition to a circular economy with the help of the EU research and innovation program *Horizontas 2020* on the European level.

The benefit of a circular economy for the main stakeholders can be distinguished into three parts: the benefit for the economy, the benefit for the companies, and the benefit for the customers and consumers. The benefit for the economy manifests in the savings on net materials when the companies' bills for the materials and the warranty risks are decreased and the costs for the consumers are reduced (on the final or multiple use products). Another benefit for the economy manifests in the volatility adjustment and the reduction of supply risk, which, in the companies, is revealed by the enhanced consumer interaction and loyalty, and the consumers gain the advantage of a greater choice and comfort (World Economic Forum 2014). The following economic benefit of the circular economy is the potential benefit of employment, which, in the companies, is observed as a reduction of the complexity of the products and a more easily manageable living cycle in companies, and its advantage for the consumers manifests in the potential secondary products usage when the latter perform more than just the basic functions. Two more benefits of the circular economy are named as reduced external impacts and long-term economic resilience (Centre for Economics and Business Research 2014). Typical circular economy activities include practices which are more oriented towards the company's internal practices and are related to internal environmental management (IEM), eco-design (ECO), and the investment recovery.

The concept of sustainable development is widely known, and the countries have prepared and confirmed their sustainable development strategies; thus, businessmen understand the benefit provided from sustainable business very well. This is useful not only for the environment but also for the company's image in order to create a positive public attitude to their activity. Reasons to transform the business into sustainable may vary: it is promoted by a number of sectors defining the laws, the international directive, the possibility to exploit the friendliness to the environment for marketing purposes, the greater opportunities to consolidate oneself in the markets where ecology is extremely important, and a wider variety of business financing and support programs and opportunities to successfully participate in them.

Each country's government seeks to protect consumers and society by controlling the production of harmful products, industrial and consumer usage of harmful products, ensuring that all consumers will have the opportunity to assess the composition of ecological products. In other words, to regulate the amount of waste which is generated by companies and organizations, each government establishes the legal system. The majority of harmful and by-products is controlled by various environmental licenses. As a result, it influences companies' behavior in creating green products or services.

On April 16, 2015, Seimas of the Republic of Lithuania confirmed the new National Environmental Strategy, which establishes the priority areas of an environmental protection policy, long-term objectives up to 2030, and the Lithuanian Environment Vision until 2050 (project No. XIIP-2686(2). The strategic objective of this document is to achieve an environment in Lithuania that would be healthy, clean, safe, and sustainably fulfilling the social, economic, and environmental needs; here, the environmental issues are also emphasized (National Environmental Strategy 2015). Four environmental priority areas are formulated in the strategy: sustainable use of natural resources and waste management; environmental quality improvement; ecosystem stability preservation; and climate change mitigation and adaptation to climate change induced environmental changes. This strategy directly contributes to the promotion of economic progress, Lithuanian "green" economic development. It will help the business to plan their activities, long-term investments, and the deployment of breakthrough technologies in a more purposeful, clear, and rational way.

The sustainable or ecological business is a good means for the company to improve its image and reputation in Lithuania since, by declaring a "green" policy, the company shows its responsible attitude to the environment and contributes to the Lithuanian sustainable development objectives. However, it is very important that the declared policy would be implemented in reality, as only then the company can expect not only the positive opinion of society but also financial benefits. "Green" technology increases the opportunities to pursue funding through various business support projects and a friendly attitude to the environment allows the company to consolidate in the markets where ecology is very important. Environmentally friendly products or services are popular and fashionable not only in internal but also in foreign markets, therefore the clients give them priority.

One of the most important features in sustainable business development and maintenance is the ability to reinvest received revenues in the company which has a number of existing or potential customers (Batista and Francisco 2018). As the environment and business are closely related, the long-term business success depends on how sustainably the organization is able to integrate into the environment and how much it corresponds to the expectations of the society and takes into consideration its problems. Earlier, the most important aim of the business was to earn the biggest possible profit. However, the modern competitive business must correspond with a number of other criteria—for instance, the environment preservation, wider social interests, greater responsibility to the local communities etc.—because this leads businesses to success. Thus, the natural and social environment and the organization are closely related and long-term success of organizations depends on how sustainably it integrates into these environments and meets the expectations of the society.

Summarizing various definitions of "sustainable" organizations, it can be stated that first a sustainable organization takes into account the person and his needs (no matter whether it is an employee, a customer, or a party of any other interested organizations); it also emphasizes the importance of natural environment and preserves natural resources through sustainable innovation, and new resource-efficient and environmentally friendly technologies and production methods. In the next section, we will discuss the principles of sustainable development of the organization and the potential principles of sustainable management applied in order to ensure the development of and the transformation to sustainable by the organization.

## 1.5 SUSTAINABLE DEVELOPMENT OF THE ORGANIZATION AND SUSTAINABLE MANAGEMENT MODELS

In an organization, the sustainability may occur in two ways: it may be built on the foundation of the sustainable development from the very establishment, or the traditional enterprise has to change oneself to become sustainable. An organization which decides to shift to the direction of sustainability can use different types of management measures that may help to improve environmental activities and working conditions, to increase the efficiency of used resources, to reduce costs, to find additional sources of income, and to become competitive in the market (Batista and

Francisco 2018). The sustainable organization is not a spontaneous phenomenon. To make the organization sustainable, the following conditions must be determined: to identify the factors of sustainability initiatives in the organization, the management system which promotes and supports them, and to develop and improve them continuously (Atkociuniene and Radiunaite 2011).

The process of transforming into a sustainable organization may be as follows (Figure 1.3): in the first stage, the concept of sustainable organization is defined and the decision as to whether to become such an organization is made. Having concluded that these changes are suitable, in the second stage, the means to achieve it are set out. After understanding the aim and what changes are needed, corresponding actions are taken. In the next stage, the achieved results are evaluated. During this phase, it is decided what will show the sustainability of the company and what will assess the results.

To become sustainable, organizations can choose the following ways: to produce goods and provide services that are in demand during a certain period of time (to fulfil public expectations and needs), to control and improve the quality of the product, to reduce manufacturing costs, to make the product attractive and to display its uniqueness, and to produce environmentally friendly products (Bagdoniene et al. 2009). The modern consumer relates the trademark directly with the seller of the service or product; therefore, ecological products reveal to the consumers the seller's concern for the environment and public welfare.

The other way to ensure the sustainability of the organization is the introduction of quality, environmental quality, accountability, social responsibility, etc. standards. ISO (the International Organization for Standardization) consists of a group of international standards that ascertain widely accepted and understood requirements and recommendations for quality management systems. The purpose of this part of the requirements is to regulate how management actions will be identified and controlled. The ISO system links the internal and external factors of the organization, which include the measures that help to change the management procedures, to reduce cost, and to increase efficiency (Simanskiene and Pauzuoliene 2011; Marimon, Heras, and Casadesu 2009). To ensure sustainable development of an organization, more than one standard can be used (See Table 1.3).

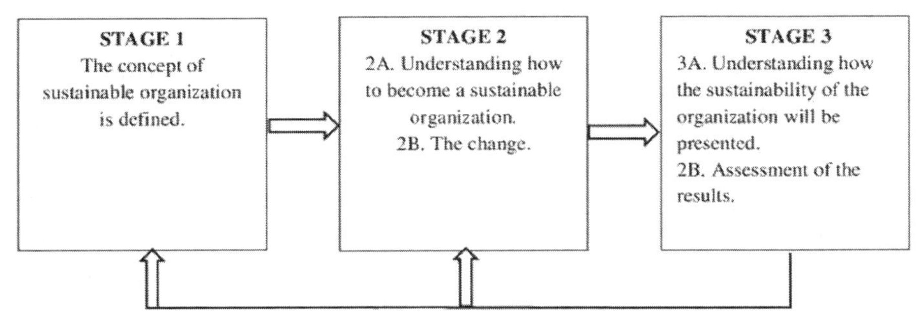


**FIGURE 1.3** The process of transforming into a sustainable organization. (Source: created by authors based on Ciegis and Grunda (2007)).

The strategic management model for transforming an organization into a sustainable business can be presented as a process of planning, changing, and evaluating (Ciegis and Grunda 2007). The planning stage consists of six steps, which start with learning and end with the objectives set by the company's sustainable development. At the planning stage, the staff's discussion on sustainable enterprise concepts, sustainable business principles, as well as the measures

**TABLE 1.3**
**Standards for Implementation of Sustainability Principles in Organizations**

| Standard | Description |
|---|---|
| ISO 9001 | The main quality management system standard consists of 20 items. Includes development, defect removal, etc. with regard to quality-related aspects. |
| ISO 14000 | The standard indicates the environmental management system requirements and specifies the implementation of the system. The standard is defined as the global management system which describes the organizational structure, planning activities, responsibilities, practices, procedures, processes, and resources by preparing, analyzing, and maintaining the efficient environmental policy of the organization. |
| EMAS | It provides the opportunity for companies to demonstrate that they operate in concordance with high environmental standards. In essence, it coincides with the ISO 14001 standard but includes stricter demands to improve the environmental situation. |
| ISO 50001 | This energy management standard is particularly intended to control energy costs and to reduce greenhouse gas emissions. The energy management standard provides structured and comprehensive energy efficiency improvement methods, i.e., this standard gives the requirements of how to create, install, maintain, and improve the management system and to continuously observe and reduce energy consumption. |
| BS OHSAS 18001 | This is the safety and health management family standard issued by the British Standards Institution. The introduction of this standard demonstrates the organization's responsible approach to safety at work. |
| SA 8000 | The social responsibility standard ensures the transparent and verifiable company business certification standards in nine fundamental areas. This standard contains the requirements for child and forced labor prevention, the freedom of workers, non-governmental organizations and collective bargaining, occupational health and its protection, non-discrimination, penalty practices, working hours, remuneration for work, and management systems. |
| ISO 26000 | This is the social responsibility guidance standard. The standard provides guidance to organizations on how ethics can contribute to public welfare. This standard is associated with the protection of the environment. |
| AA 1000 | This is the accountability standard, ensuring sustainability. It provides recommendations based on reliability and quality requirements. |
| ISO 26000 | This standard aims to create added value by structuring and presenting universally accepted principles in the area of corporate social responsibility. The basic principles include such areas as environmental protection, human rights, recruitment and employment practices, the leadership of the organization, transparent activities, community involvement and consumers. |

*Source:* created by authors.

and their applicability in a particular organization is encouraged, standards and systems are debated, and an attempt is made to show the results revealed in order to achieve sustainability. This phase includes the company's situation analysis, which is carried out by concentrating on the imbalanced business areas. One of the steps involves creating a vision for a sustainable organization and the setting out of clear and easily measurable objectives.

The changing phase is subdivided into three steps: planning, priorities, and measures. The objectives set out in the first phase include the specific measures which the company applies in order to modify a particular product or process. The second step involves prioritizing certain measures. This clarifies the measures which will most likely help to achieve the objective of transforming a company into a sustainable one. After granting the priorities, measures are implemented. During the changing phase, the company's structure may be reformed, the employees sorted out or substituted, the consultant assistance applied, and the suppliers, the production processes, and the production designs replaced with others. The buying process, marketing, logistics activity, etc. play a fundamental role in this process. The results of the evaluation phase include a comparison between the objectives set during the first phase and the company's achieved results. During this phase, it is tested whether the planned certificates are received, whether the products and processes comply with the planned standards, whether necessary reports are being prepared, and whether the company has been connected with the organizations and the certain environmental trademarks are obtained.

In order to transform an organization into a sustainable organization, it is important to take appropriate, timely measures and to apply the most suitable methods for the organization (Atkociuniene and Radiunaite 2011). There are various means available to achieve the objectives of transforming the organization into a sustainable one (see Table 1.4).

*The Natural Step* is used as a measure for the company to achieve sustainability. However, it does not accurately provide the company with specific instructions which can lead to organizational sustainability. In contrast, business partners are encouraged to create their own actions and approaches tailored to each situation (Nathan 2018). *The Natural Step* defines a sustainable company through its links with the availability and the use of resources by applying a "resource funnel" model and encouraging the watching of the current situation from a future perspective, i.e., by applying the *back casting* method. This method was developed by *The Natural Step*, founded in 1989 in Sweden by K. H. Robert. The organization's mission is to promote global consistency by guiding companies and governments in the environmentally, socially, and economically sustainable direction. *The Natural Step* works together with one of the world's largest resource users to create solutions, new models, and measures aimed at a sustainable future.

*The GEMI sustainable development measures for business* are the Global environmental management initiative GEMI, established in 1990, in the USA. Its main purpose is the worldwide deployment of the eco-management principles to promote business ethics based on environmentally safe management and the concept of sustainable development, the proposition of different measures to help the companies to achieve sustainable development (Burlea-Schiopoiu 2013).

**TABLE 1.4**
**Measures of Transforming the Organization into a Sustainable Organization**

| | |
|---|---|
| 1) *The Natural Step* as a measure to achieve sustainability | 13) Green purchasing |
| 2) GEMI sustainable development measures for business | 14) Contracting for results |
| 3) Triple bottom line accounting | 15) Pollution prevention |
| 4) Industrial ecology | 16) Zero-emission |
| 5) Cleaner production | 17) Quality management |
| 6) Eco-design | 18) Global cost assessment |
| 7) Product life-cycle evaluation | 19) Eco-marketing |
| 8) Eco-efficiency | 20) Ecological logistics |
| 9) Energy efficiency | 21) Sustainable ecologically balanced system of indicators |
| 10) Environment-friendly production | 22) Environmental management system |
| 11) 4R—reduction, re-use, recycling, extraction | 23) Sustainable leadership principles in the installation |
| 12) Factor 4 and Factor 10 | |

*Source:* created by authors.

*Triple bottom line accounting* refers to the non-limitation to the traditional generation of profit-and-loss account of an enterprise by expanding the reports and including environmental and social analysis of the results. In accordance with the triple bottom line concept, organization sustainability focuses on the harmonization of the environmental, social, and economic human activities dimensions in such a way that economic growth would be pursued with the consideration to the long-term environmental and social integration.

*Industrial ecology* promotes the closed production cycle model, in which the resources are used in efficient eco-cycles. Industrial ecology is the integrated systemic approach to energy, materials and capital efficiency in industrial ecosystems (Pongrácz 2006).

*Cleaner production* can be defined as a preventive integrated environmental management strategy, which must be applied permanently to the manufacturing processes, services, and products throughout their life-cycle in order to reduce the negative impacts on humans and the natural environment.

*Eco-design*, or updated design (also called environmental design, environmentally sustainable design, etc.), is the concept of physical objects, created environment, and service design, which is implemented on the basis of economic, social, and environmental safety principles. The goal of eco-design is to completely eliminate adverse effects on the environment using professional and sensitive to the environment design methods (Navajas, Uriarte, and Gandía 2017). Such design projects are based on a minimum consumption of non-renewable resources and minimum environmental impact, in order to connect people with living nature. Objects of this area range from the microspheres, the small objects of everyday use, to macrospheres, buildings, cities, and the landscape. This is a concept that can be applied to architecture fields, landscape architecture, urban development, urban planning, designing, graphic design, industrial design, interior design, and fashion design, in order to ensure more environmentally friendly production processes and products and to maintain the desired price of the product and performance characteristics.

*Product life-cycle assessment* is the storage and analysis of the data on costs, output, and of potential environmental impacts throughout a product life-cycle (Gehin, Zwolinski, and Brissaud 2008). As a result of such assessment, the company can clearly see which product areas need improvement to reduce the overall negative environmental impact of the product. The product life-cycle can be divided into four stages: product stage (raw material production, supply, transportation); development phase (production, distribution, transportation, installation); operational phase (product use, maintenance, repair, replacement, refurbishment); and end of life stage (dismantling, demolition, transportation, recycling or disposal).

*Eco-efficiency* is reached by supplying competitively priced goods and services that meet the needs of people and provide them with quality of life, by progressively reducing ecological impact and resource intensity of the goods in the life-cycle to the level which the ground capacity can manage. The following indicators characterizing the efficiency of business enterprises are distinguished: technological (technical), economic, allocative (distribution), and ecological. Technological efficiency may be understood as follows: a production is considered to be technologically efficient if the same production technology produces the same output at a lower cost, i.e., less physical, financial, labor and energy resources are used or with the same resources the production volume is increased. Economic efficiency is the achievement of better results at lower costs (United Nations 2009). Allocative or distribution efficiency is the most suitable trade mix production at the lowest cost or optimal allocation of economic resources. This efficiency also means that the best available combination of resources is used. Eco-efficiency is the optimal alignment of all technological processes, using the most modern technological equipment and ecologically clean technologies. Each of the efficiency indicators above has a critical influence on the overall efficiency of the company.

*Energy efficiency* is one of the most important eco-efficiency indicators. Higher energy efficiency is ensured by the use of such products which provide the same service, only by using less energy.

*Environment-friendly manufacturing* is responsible for new product manufacturing methods from conceptual design to final delivery and ultimately for meeting environmental standards and requirements at the end of consumption. Environment-friendly manufacturing is ensured through the application of preventive environmental strategies to improve environmental performance, reducing the costs and risks to humans and the environment. In the manufacturing processes, it includes energy and raw materials conservation, hazardous material disposal, and reduction of the hazard and the quantity of the total emissions before leaving the process. In the area of products, it aims to reduce the impact the product has on the environment throughout all of its life-cycles, i.e., from raw material extraction to the final product depositing. In the area of services, it involves the preventive strategies including design, investment-free measures, and more environment-friendly promotion of choice of raw materials or products.

*4R—reduction, re-use, recycling, extraction* encourages the transition from waste management to waste prevention. Its essence is to achieve a significant reduction of the waste stream in each organization by applying repeated methods of increasing the efficiency of raw material utilization, recycling, production, and extraction.

*"Factor 4" and "Factor 10"* are the concepts which state that it is necessary to seek fourfold, or even tenfold, more efficient use of resources. Fourfold effective use of resources arises from twice as much reduced costs and by achieving a doubly increased efficiency. Factor 10 developers claim that, in the long term, in developed countries, the use of resources will have to be reduced tenfold if the intention is to achieve sustainable development performance. It is argued that at the global level consumption has to be reduced twice, but the maximum consumption must be decreased in the countries which are currently consuming the most resources.

*Green purchasing* is carried out when the choice of the goods or services is not only (or even not) the price and quality, but also the environmental impact of the production or consumption process of the purchased product. In accordance with the green procurement principles, companies must perform an environmental assessment of the goods being procured at various stages of use and the entire item life-cycle.

*Contracting for results* is used as a measure to obtain financing for investments to improve energy efficiency.

*Pollution prevention* focuses not on the means to reduce pollution, but on how to avoid it. Pollution prevention principles are best to follow when creating a new product or process because pollution prevention used on the existing goods requires their modification, and this may demand changes in the components, design, or manufacturing processes.

*Zero-emission* is the concept that aims to achieve the maximum resource productivity and full elimination of pollution, and the very process is based on naturally occurring cycles in which there is no waste.

*Quality management and total quality environmental management* are based on two principles: prevention and inspection. The application of these systems also allows the efficient use of resources and the reduction of waste.

*Global cost assessment* is the concept whereby the essence of which is to evaluate not only the private and external costs arising, for instance, due to the pollution. Following the evaluation of the social cost, the decision-maker knows all the options, costs, and opportunities to make the best decision.

*Eco-marketing* is a specific product development, costing, promotion, and distribution without the damage to the natural environment. Eco-marketing tasks are related to the products and production methods which raise environmental problems or solve them. It can also be described as a holistic and responsible management process that identifies, anticipates, satisfies, and fulfils the stakeholder requirements for a reasonable reward, which does not have adverse effects on the well-being of the human and natural environment. The concept of sustainable marketing is also applied. It is considered as the eco-marketing sequence, a step towards the development of sustainability. It is a broader management concept which focuses on sustainable solutions to the development and realization, providing a higher net to the sustainability value, while at the same time fulfilling stakeholders' needs. Eco-marketing in the organization is related to all company activities and its philosophy. Organizations which rely on an ecological concept should direct its attention towards the three external stakeholders: consumers, society, and the environment.

*Ecological logistics* is logistics operations performed in a way that makes the environmental pressures, such as traffic jams on the roads, air pollution, fuel efficiency, and waste generation minimization, a key factor in the political decisions (McKinnon, Browne, Whiteing, and Piecyk 2015).

*Sustainable ecologically balanced system of indicators* enables monitoring of the progress towards implementing sustainable development goals set by the organization by applying these indicators.

*The environmental management system* is an established, implemented, and functioning system for the management of significant environmental aspects to ensure compliance with requirements of environmental laws and regulations.

All of these measures help to improve working conditions, to increase the material and energy efficiency, to reduce costs, to find additional sources of income, and to increase the competitiveness of organizations. The discussed measures allow the company to distinguish sources of waste or pollution and the cause of their formation, and to offer ways for the re-use, reduction, or withdrawal of these products.

*Sustainable leadership* is identified as one of the instruments to achieve operational sustainability, the concept of which is provided in detail in the monograph by L. Simanskiene and E. Zuperkiene, published in 2013. The concept of sustainable leadership was first used to address youth education and health issues (Simanskiene and Zuperkiene 2013). The importance of achieving sustainability at different levels has led to wider dissemination of this concept. The concept of sustainable leadership is used at different levels: at an individual-psychological and the physiological health level; at an organizational level, by maintaining and fostering a working environment and culture where the employee can thrive and realize their potential in pursuit of the organization's objectives, which are consistent with his personal goals; at a sociological level, by taking a responsible position in the community; and at the organic level, by protecting the environment surrounding us. Sustainability principles are related to thought, action, and knowledge changes, which may be based on: the creation of self-awareness directed towards coherence; the knowledge creation geared towards sustainability; learning to use the latest environmentally friendly systems in the organization and the principles of sustainable manufacturing. In order to create long-term organizations and to motivate employees to provide the highest level of service and to create values of organizational sustainability, conscious leaders who lead in good faith and respect ethical principles are necessary. Studies have found that leaders who seek sustainable activity are similar to other effectively employed guides, only they have additional leadership characteristics: they think in a broader way than is imposed in the organization, and they take into account the stakeholder interests. Leaders who seek the organizational sustainability must pay attention to the fact that foremost the sustainable development ideas are possible to maintain through the organizational culture, involving all interest groups (Simanskiene and Zuperkiene 2013).

Responsibilities of the sustainable organization activities pose a growing complexity of requirements to their managers: when implementing the company's strategy for sustainable development, leaders are responsible for the coordination of the goals, values, and procedures across all organizational systems and chains; they also ensure the managers' and employees' loyalty promotion and successful collaboration among all stakeholders.

According to L. Simanskiene and E. Zuperkiene, the sustainable leadership pyramid consists of three levels: the basis consists of 14 basic practical activities, six practical activities of a higher level, which are implemented on the baseline basis, and the third level includes innovation, quality. and involvement of the personnel. Practical action has an impact on five activity results: the brand and reputation, customer satisfaction, operational finance, long-term share value, and long-term value for many interest groups. Sustainable leadership primary is distinguished by the following characteristics: a clear business vision focused on sustainability; long-term rather than short-term objectives; leaders taking responsibility not only for the work of their own and of the working group but also for each individual, group, organization, and society; the organizational culture is strong, focused into a coherent organization development; not only individual efforts, but also joint efforts are focused to help each other to achieve performance; managers have a high degree of human confidence and goodwill; the results of the work are characterized by team members' joint effort in synergy; changes in the organization are initiated not as the destruction of old and the creation of new, but as the re-using, giving, and joining of resources that already exist within the organization; cooperation between members of the organization is ongoing and continued; the attention is focused not at the group work, but at team work; quality is achieved not through control but through an organizational culture focused on sustainable development; the leaders and members of the organization are aware of what sustainability is and are guided by the principles of sustainability; employees are loyal to the organization because it meets their needs and their safety is ensured; all members of the organization are developing; the qualification of employees rises constantly, and they are cared for; innovations are installed systematically and continuously, creativity is promoted, and a sufficient amount of resources is dedicated to promoting creativity; working relationships are not formal but cooperative.

Leaders seeking sustainable development are able to agree and unite different approaches. They give employees more power and have a great influence on them. In addition, they are empathetic, they care for the staff, and they maintain good emotional ties with them. If such leaders need help, the staff are always willing to help them. This definition of sustainable leadership is suitable for all areas of the organization's activities. The essence of sustainable leadership can be described as follows: sustainable leadership includes everything that spreads and lasts, without harming anyone and having a positive impact on everything that surrounds us now and in the future (Hargreaves and Fink 2004; Ferdig 2007). Personal input is needed to create organizational sustainability: it is necessary to change the mind-set of each individual involved. Thus, self-consciousness begins with the change in the individual, then moves to other levels: group, organization, and society.

Thus, it can be stated that sustainable leadership is a leadership when there is a sense of responsibility for individuals, groups, organizations, community, assessing the environmental, social and economic principles of sustainable development in the context of a group, organization, community, promoting sustainable development ideas, promoting sustainable development based on teaching and learning, and human expression (Morsing and Oswald 2009).

Sustainable leadership complements the concept of leadership by incorporating new variables. It is no longer the elementary leadership activities, when in order to achieve the desired objectives, the behavior of others, both individually and in groups, is influenced. In the case of sustainable leadership, a manager seeking to influence the behavior of others has to assess all important economic, environmental, and social aspects and only then make decisions. The result of sustainable management is associated with the comprehensive benefits not only for specific members of the organization or for the organization as a whole, but also for the society, having regard to its long-term needs: to maintain coherence with the environment, both economic, social, and ecological aspects. Various areas of activities must be included in order to lead in a sustainable way. Sustainable leadership can be analyzed at three levels: individual, group (team), and organization. For the sake of leadership sustainability, it is important to properly assess the leader's personality and creativity, to bring together a harmonious team by assessing harmonious staff relations and expertise, fostering loyalty and organizational culture of an employee, assessing and developing the corporate image and corporate social responsibility. Although it is clear that each area has an effect on each other and everything is interconnected, relatively it can be attributed to the individual and group (team), and organization levels. This is depicted in Figure 1.4.

The purpose of sustainable leadership is to focus the organization and its members towards sustainable development, to carry out socially responsible practices, and to apply socially responsible corporate practices. Sustainable leadership is not easy to implement. It not only depends on manual skills and determination of

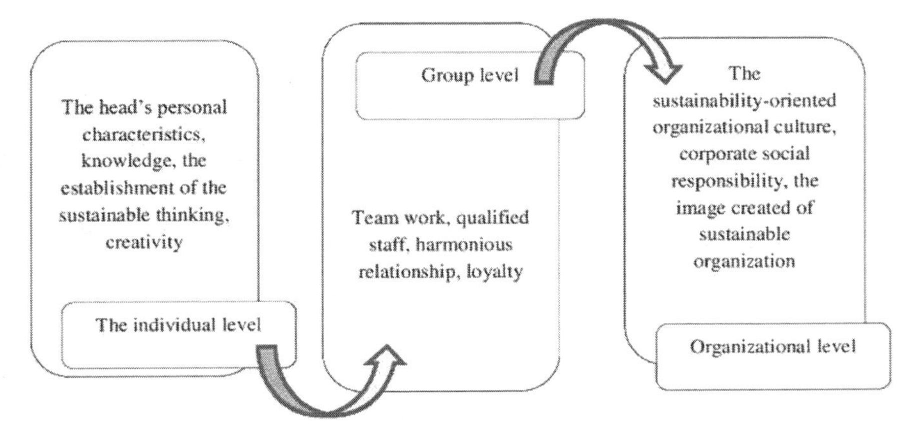


**FIGURE 1.4** The essence of sustainable leadership of estimating the three levels. (Source: created by authors.)

the leader, but permanent attention and efforts in order to progress must also be maintained; however, most of the leaders and their followers often burn out, and other managers are not always able to continue the work and maintain the results achieved. The biggest benefit is perceived as the result of organization activities, and it is beneficial not only for the organization itself in the form of the final product or service but also for the community and the environment with which and in which the organization operates. Sustainable management aims to avoid any negative consequences on economic and social development. Sustainable leadership fosters the highest values, generally for the public, which are based on the performance of the organization's development and the management of anticipated risks.

In conclusion, it can be stated that those organizations that are built on the foundations of sustainable development have a better understanding of sustainability and its significance. Businesses that want to reorient themselves experience more difficulty since they may need to change their whole working methodology. In order to achieve organizational sustainability, first, the whole process needs to be planned; only after that is it installed, without forgetting to evaluate the achieved results. Whatever method is chosen in the organization, the most important is to understand the meaning of sustainability and what results it may give.

However, currently, there is a problem of creating the image of a sustainable organization. It is a relatively new phenomenon called "greenwashing." It is the abuse of the ecological watchwords, manipulation by false information, and deceiving the consumer about the apparent product ecology (Gräuler and Teuteberg 2014). On the one hand, the fact that such effects exist, and, unfortunately, occur quite often, shows the lack of consumer knowledge about organic production standards, product composition, and ethical production principles; on the other hand, the "greenwashing" phenomenon suggests that green criteria in society are treated as added product value; therefore, it is attractive to the manufacturer who wants to gain a competitive advantage. Non-governmental environmental organizations, like Greenpeace, are actively involved in the fight against "greenwashing." Online websites aiming to prevent the "greenwashing" are created, books about the "greenwashing" are published. Some examples of "greenwashing" are: vague, abstract, and ambiguous statements, empty watchwords, the phrases "natural," "environmentally friendly," and "good for the planet," which are not justified by the relevant facts, or the manufacturer claims that the consumer can "save the planet" by purchasing their product. Another example is a visual or graphic presentation. Most of the green goods are used in the color gamut of the package to make an ecological impression and it is stated that the product does not have any harmful effects on the environment. Usually, these claims are not approved by the third independent and trusted sides, there are no marking/certification signs, no comprehensive product composition is provided, or "green" goods or services represent only a small fraction of all the company's activity area. The debate on sustainable development has attracted the focus of easy-profit companies; even large companies like General Motors, Shell, and Mitsubishi have been tempted by eco-manipulation. Such dirty games are very dangerous for the reputation of the business and bring only a short-term effect, and when the truth comes out, the company risks irreversibly losing the trust of consumers. Therefore, it is better to be an honest non-organic manufacturer or seller, rather than to simulate ecological and social responsibility. Recently, a number of goods have appeared in Lithuania, especially in the group of food products, with expressions such as "healthy" and "green," and the potential buyers are manipulated by the use of colors—the image of the organic product is created unreasonably, regardless of its production technologies, composite materials, etc. Sometimes companies that pollute the environment in their production processes promote environmental initiatives. This is another example of an artificial "green" image of the organization.

Organizational culture plays an important role in the company's sustainable development and protects from temptations of green brainwashing. The high-culture organizations do not rely on short-term benefits in achieving goals and do not engage in manipulation. Advanced management methods and leadership principles are primarily applied in organizational culture development. The next section deals with the concept of organizational culture and its importance for the company's transformation into a sustainable organization.

## 1.6 THE IMPORTANCE OF THE ORGANIZATION'S CULTURE FOR TRANSFORMATION INTO A SUSTAINABLE ORGANIZATION

Currently, the organizational culture plays an important role in addressing the quality and improvement of services provided to the public. An organization which has a scale of values and follows its established standards of conduct, traditions, and appropriate views towards the customer usually provides quality services and produces high-quality goods. The examination of the internal climate of the organization and its culture particularly actively began in the 1980s. In addition, the impact of cultural values on the efficiency and success of the organization's activities was also heavily debated. Therefore, many researchers analyze organizational processes: employees' governance as well as actions which had an impact on the formation of their cultural provisions and values.

Lithuanian scientists do not agree in discussing what term should be used in terms of an organization's culture; whether the prevailing culture in an organization is organizational or the culture of the organization (Juodaityte 2003; Zakarevičius 2004; Pociute 2005). It can be said that both the concept of an organization's culture and organizational culture emphasize the same characteristics since, in their insights, which define the concept of an organization's culture, the essential elements revealing an organization's culture are found. In their works, L. Simanskiene and E. Zuperkiene examined both organizational and organizations' cultures and claimed that in order to define any culture of organization, two different concepts must be employed: "organizational culture," when the management theoretically knows about this culture and practically forms and nurtures it, and "organization's culture," when the management does not know this concept and consciously does not form appropriate values for employees (Simanskiene and Zuperkiene 2013). So, the diverse perception of authors differs mainly because they use two terms which are similar, but their meanings are slightly different.

According to the statements of various scientists, it can be claimed that an organization is a component of the socio-economic system; therefore, in order to describe conceptions, such terms as "national culture," and "human culture" are used. Hence, by analogy "organization's (specific group of people) culture is used" (Leithy 2017). This term is the most suitable because such a term as "human culture" does not reflect the importance of organization and can be used in any other system. It is believed that these two different concepts have emerged due to inappropriate translation from the English language and also because these concepts analyze the relationship between the organization and its culture, as it makes the same impact on the organization's development no matter how the term is defined.

According to some authors, an organization's culture defines provisions, beliefs, expectations, attributes, and habits of its members as a whole (Rollins and Roberts 1998; Biswas 2009; Blackwell 2006). It shows the company's group consciousness, which leads to its reaction to the processes taking place inside and outside of the organization and which determines their behavior. It can be stated that an organization's culture is strongly linked to its employees. It reflects their behavior, thinking, and external appearance (structure, symbols, heroes, etc.), which formed during communication among employees and with the external environment. First of all, it manifests within provisions, beliefs, and values of employees.

Thus, an organization's culture is the system of essential values, which are specific to the organization and its members. The latter are guided by these values in their daily activities. Such a system manifests by traditions, rituals, stories, and myths. Here, it is very important to highlight the significance of values, which as well as the standards of conduct, approach of a manager and employees' commitments strengthen the already formed culture of the organization. Hence, organizational culture mostly manifests through each member of the organization and their daily, values-based activities (Kalaiarasi and Sethuram 2017).

Different cultures dominate in organizations depending on their nature. Organizations which are characterized by a distinctive organization's culture are unique and increase employees' attachment to the organization. The main cultural characteristics of organization correspond to the natural development or national culture, slow changes, universality, and behavioral standardization. All organizations are different because they develop and change differently; also, they are characterized by different behavior based on the activities the organization carries out and the kind of employees that work there. Each organization is distinctive and, according to its uniqueness, it creates corresponding functioning conditions under which an organization's culture encourages successful operation. Then it can be called a strategically appropriate culture. It can be claimed that the most important thing is that the organization's culture should direct its employees to act in the right direction. Since the organization's culture depends on many factors, the culture cannot always remain the same. The organization's culture must change, because the environment changes and people have to deal with the new internal problems of the organization by learning and gaining new experiences (Shehri, McLaughlin, Al-Ashaab, and Hamad 2017). Thus, some organizations have time to change their cultures according to market conditions, while others apply different methods; therefore, the culture of all organizations is different.

The organizational culture is examined more widely, and here we are faced with various opinions of both Lithuanian and foreign authors. Therefore, it can be said that organizational culture is one of the company's strategic resources, which has an impact on the company's activity, efficiency, and successful business development. Furthermore, it has a lot of impact on the achievement of the organization's objectives and the successful implementation of its vision. When the organization has a goal to thrive and to meet the needs of its members, certain symbols are used and norms and values are created to reflect the attitudes of the members of the organization on how to work in order to achieve these goals (Steven 2000).

The importance of organizational culture has recently increased, and this promotes greater organizational commitment. Broadly speaking, organizational culture helps organizations to achieve their goals by affecting their activities and employees. It is defined as a system of essential values and beliefs. This system is recognized by all the employees, it is supported by the stories, myths, and heroes of the organization, and has an impact on the behavior, as well as manifests through norms, traditions, language, and symbols (Young, Yom, and Ruggiero 2011). By using certain symbols, and creating certain values and norms, the general provisions of all the members of the organization on how to work in a way so that the organization can thrive and meet the needs of the members of the organization are reflected. Thus, organizational culture primarily is a behavioral model, based on beliefs, values, and attitudes, which may have an impact on the work practices. It shows that both the organization's culture and the organizational culture include the values of the organization.

In the organizational culture, the role of the employees is particularly emphasized. Organizational culture is a model based on the basic assumptions that the group itself must invent, discover, and create learning methods on how to cope with its own problems. Organizational culture allows the organization to adapt externally by applying internal integration. In order for this model to function well, it must be used, and new members of the organization must be trained in how to correctly perceive, think, feel, and behave when solving organizational problems. Thus, organizational culture includes the employees from all hierarchical levels and is based on the long-lived history, which is implemented by significant organizational aspects—for example, its name, products, buildings, logos, and other symbols. In terms of the organization's and organizational culture, employees constitute an important part.

Organizational culture is very important for companies because it includes employees in the governance process and has an impact on all the organization's activities (Rose, Kumar, Abdullah, and Ling 2008). Organizational culture is a method of governance that can be used to improve the quality of the governance process since, due to people's commitment and loyalty to the organization, great results of work can be achieved. Thus, organizational culture is a relevant and effective management tool which has an impact on organizational activity and its results, as well as on the relationship with society and the environment. Furthermore, it plays an important role in such areas of the organization as the implementation of key initiatives, and the organization's speed, when responding to the market changes and organizational opportunities due to the variations of the business environment. Regardless of the organization activities, its culture is an integral part that has an impact on the company's governance and results.

Although organizational culture is deeply integrated into the company's governance and management processes, it does not always facilitate the governance process. It is a medium which forms and supports the distinctive environmental conditions where the governance is happening. Such medium either facilitates or complicates the governance and functioning processes of the organization and has an impact on its results, but in any case, the organizational culture must be coordinated with the governance process, considering fostered values, traditions, etc. A positive result can be obtained precisely when the organizational culture is aligned with governance actions. It is not an easy task because the organizational culture is a rather complex phenomenon, which has various forms and aspects of expression. Organizational culture is of great importance for complex corporate governance and management decisions; that is why it is so widely analyzed and discussed in management research.

The creation and changing processes of organizational culture in the desired direction of the organization are primarily based on cultural values or values that can be used to unite the efforts of all members of the organization. Upholding values is a means of achieving the goals of an organization. Such values do not necessarily have to correspond with universal cultural values. They may even oppose the universal cultural values in a certain country. First of all, values include symbols, stories, rituals, traditions, etc. Since the members of the organization are very important to its culture, core values, which are firmly embedded and perceived, strongly unite the employees of the organization. Hence, values in the organization's culture have great meaning and the success of the organization's evolution and cultural development depends on them (Kenny and Reedy 2007).

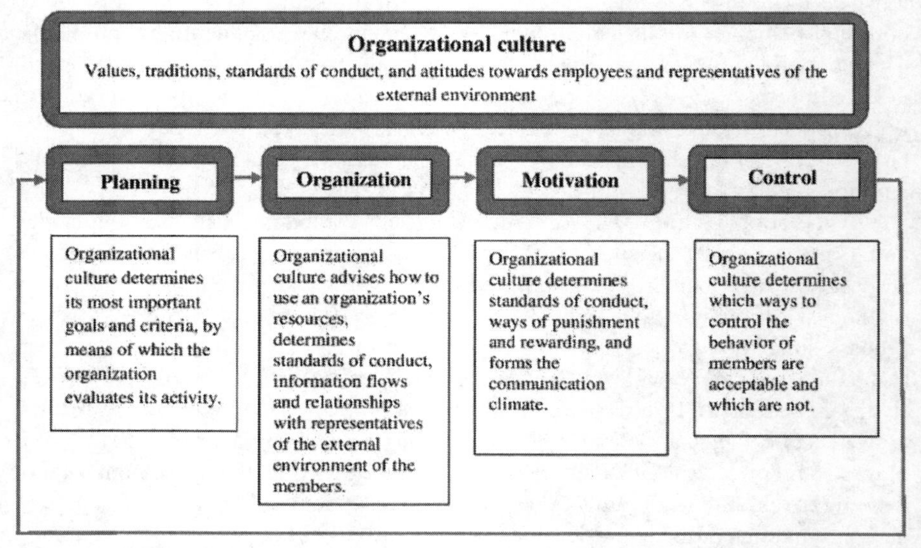


**FIGURE 1.5** The model of the role of organizational culture in the management process. (Source: created by authors.)

The role of organizational culture in the organization's management process is very important. Figure 1.5 shows values, traditions, standards of conduct, and attitudes towards the employees and the representatives of the external environment divided according to four interrelated components: planning, organization, motivation, and control. In the part of planning, organizational culture determines its most important goals and criteria, by means of which the organization evaluates its activity. In the part of the organization, organizational culture advises how to use the organization's resources, determines standards of conduct of members, etc. In the part of the motivation, organizational culture also has a role to play, as it primarily forms the communication climate within the organization. Organizational culture also determines if the behavior of the organization's members is acceptable or not. This ensures control within the organization. Such a model shows that organizational culture has an impact on the stages of management processes, which are closely interrelated.

Organizations are different in terms of the organization's cultures and organizational cultures. Organizational culture can be called artificial, and deliberately developed culture. If the structure of organizational culture corresponds to the structure of the organization's culture, their content still differs. Even if the content of the components may be similar, they are still different, and this distinguishes organizations from one another. Characteristics of organizational culture form the content of an organization's management and become its distinctive features. According to these features, organizations position themselves, and become recognizable and interpretable by staff, users, customers, and other people. Organizational culture is related to continuous improvement of activities and changes through technological and organizational management structures, the creation of new products, new management methods, and other activities. These actions help organizations to maintain competitiveness when some organizations are better than others. In their competition, organizations use various methods to improve the activities of their organization, as well as to adjust their organizational culture.

To conclude, it can be argued that organizational culture is created artificially; having a certain vision and goals and being aware of the mission of the organization, and the culture of the organization, is a well-established phenomenon. It is important to distinguish the following organizational culture elements: values, provisions, norms, policy, vision, mission, ideology, philosophy, heroes and stories, rituals and ceremonies, material symbols, language, and cultural communication network.

Since companies aim to maximize productivity as well as cost and waste reduction, it is important for organizations to be able to respond in a timely manner not only to external developments but also to active market players that create markets that are capable of exploiting benefits because it is the only way to remain competitive. Organizations aim to make employees work willingly and productively, show initiative, take care of customers, and work actively with other companies in a similar profile region, but we rarely think about the fact that this cannot be achieved without a well-developed organizational culture, which is one of the tools to achieve organizational sustainability and social responsibility (Martin 2002).

The sustainability of a company can be ensured by implementing certain forms of organizational behavior (including public influence, environmental impact, culture, and finance), using management tools and available human resources (Linnenluecke and Griffiths 2010). The company's sustainability or the organization's transformation into a sustainable company could ensure the application of leadership principle because leadership is an effective tool to form the organization's culture. Formed organizational culture encourages employees to work productively and can be viewed as the best motivation; however, the creation of such good organizational culture in a company is quite difficult, since it is not enough to create it, as it requires constant maintenance, fostering as well as reaction to changes happening in organization and environment. The creation and fostering of organizational culture, as well as the compatibility of values with the organization's employees, is a difficult problem. The members of a strong organization's culture unanimously agree on its goals and such unity creates harmony, loyalty, and dedication to the organization. It should be noted that it is very important that the values of both the organization and its members coincide, and the companies must attract such employees; if new employees do not adapt, they should leave the company.

It can be argued that an organization's culture helps to transform the organization into a sustainable organization. When the values of the organization coincide with the values fostered by its employees, the employees work willingly, productively, strive for the company, and are proactive; thus, a strong organization's culture is achieved. When the organization's culture is strong, companies on the market are competitive and reach a wide range of objectives, such as higher profits, satisfied customers, productivity, declining costs, and reduction of pollution and negative impact on the environment. When an organization reaches these objectives, it becomes sustainable and can spread sustainability by constantly monitoring and evaluating whether a higher level is achieved. The following components are necessary for the formation of an organization's culture and formulation of values and leadership of cultural changes in the organization: to lead the process, to appoint a leader of change, to identify threats that stimulate change, to carry out transition rituals, to organize intensive training, to provide new direction signs, and to emphasize employment security. As you can see from these components, the essential one is the leader, who manages transformation processes. Thus, only through leadership can the new organization's culture be created and transformed into a harmonious one.

High-culture organizations recognize and foster values of sustainable development and strive for sustainable development objectives. These organizations recognize the principles of social responsibility and their importance and seek to position themselves as socially responsible companies since this, in particular, meets their values. As business can make a significant contribution to the development of an equal, healthy, and vibrant society, and the concept of corporate social responsibility allows us to justify the importance of sustainable business development for achieving the country's sustainable development objectives, the following section presents the main insights of scientists and practitioners on corporate social responsibility and its role in society.

## 1.7 CORPORATE SOCIAL RESPONSIBILITY IN THE CONTEXT OF SUSTAINABLE DEVELOPMENT

Since the concept of sustainable development is a key concept for the development of modern society, its realization faces a number of obstacles, primarily due to the inability of the market to solve environmental, public goods, information limitations, social inequality, and other problems. A free market in itself cannot ensure the achievement of social and environmental objectives of sustainable development. Therefore, government intervention in the market is necessary in order to evade or fix these setbacks. However, government regulation alone is unable to solve these market problems, especially on a global scale.

The EU's *Shared Responsibility* policy sets out new market approaches in order to reduce and regulate the negative impact on the environment and to spread information widely. The core of the policy consists of the principle of global responsibility and the support of fields in voluntary activities, and social protection. In their policies, international organizations and governments emphasize that traditional environmental control mechanisms and government regulation based on strict administrative methods and coercion can no longer ensure a stable situation in the environment, environmental protection, and social life (Korhonen 2003). It is important for companies themselves to continuously improve their performance in environmental protection and working conditions. For this purpose, alongside mandatory environmental and labor relations regulatory requirements, which are defined by laws and other legal acts, new means are being created in order to encourage companies to freely pursue better environmental and social standards. Voluntary environmental and social activities are becoming an important factor in improving the image and value of a company. In addition, this activity is economically viable and guarantees an increase in the company's competitiveness.

The government, international organizations, and the public became interested in the voluntary environmental activities of companies relatively recently, only in the late 1980s. These corporate initiatives complement mandatory environmental and social requirements regulated by public authorities.

Thus, voluntary business initiatives, the origins of which lie in the concept of a social contract, play an increasingly important role in the global sustainable development policy. The role of business in the economy is crucial; therefore, such voluntary initiatives at both global and micro levels, such as the United Nations Global Compact, the Caux Round Table, the development of socially responsible investments, and the implementation of social and environmental management standards, humanized workplaces, and strategic leadership for sustainable development, are the key tools for enabling sustainable development and implementing these principles at local, regional, national, and global levels (Jenkins 2005).

It is clear that the vision of sustainable development for business organizations poses new challenges, which are summarized in the concept of enterprise (corporate) social responsibility. Socially responsible companies have to analyze and properly evaluate the social consequences of their activities and their environmental impact.

The World Business Council for Sustainable Development provides the following definition of CSR (corporate social responsibility): it is an ethical behavior of the company towards the community, which consists not only of shareholders but of a much wider circle of stakeholders with legitimate, business-related interests.

The idea of corporate social responsibility was constructed and published in the European Union Commission Document Promoting a European Framework for Corporate Social Responsibility, COM (2001) 366, released in July of 2001, and presented by the European Commission in the Green Paper. CSR is defined as a concept in which companies voluntarily integrate social as well as environmental issues, and collaborate with stakeholders in order to ensure a sustainable and successful business. Organizations voluntarily decide to contribute to the improvement of the quality of life in the community and the preservation of an intact environment.

Thus, it is clear that CSR is perhaps one of the three most important elements of sustainable development; it is also a driving force that promotes socially acceptable activities. Corporate social responsibility is based on three core elements: *economic, environmental*, and *social*. The concept of corporate social responsibility is fully consistent with the concept of sustainable business or sustainable organization. Every socially responsible company can be considered as sustainable and directly contributing to the goals of sustainable development at the micro level (Jamali and Mirshak 2007).

The UN initiated the Global Compact, which is a key tool for developing corporate social responsibility. The Global Compact is a global initiative presented by UN Secretary-General Kofi Anan during the World Economic Forum in 1999. He invited business leaders to join the initiative on the basis of ten universal principles concerning human rights, workers' rights, environmental protection, and anti-corruption. The aim of this agreement is to encourage businesses to carry out their business responsibly, not to harm the environment, the community, or other business, and, by means of joint efforts with the UN, to encourage governments and the non-governmental sector to address specific social and environmental problems in society and in this way contribute to socially inclusive economic growth. The initiative provides the opportunity for companies and organizations to share knowledge, experience, and innovations on how to improve the business strategy and corporate image, as well as to improve the risk management strategy.

Currently, there are several tens of thousands of companies and international labor and civil society organizations from around the world that are involved in the Global Compact. The project launched by Kofi Anan is considered to be the largest voluntary corporate citizenship initiative. After becoming a member of the Global Compact, the name of the company or organization and its social responsibility practices are included in the list of participants published on the internet. This in itself is a big advertisement for the entire organization and raises the image of its brand in the eyes of the consumer. It should be emphasized that the Global Agreement is a voluntary corporate citizenship initiative with two binding objectives:

- to make the Global Agreement and its principles a part of business strategies;
- to promote cooperation and partnership that support UN goals.

In order to achieve these goals, the Global Compact offers support for the organization of political dialogues, learning, and partnership projects. A Global Agreement is not made to oversee, monitor, or measure the behavior and actions of companies. The Global Agreement is based on public accountability, transparency, and business interest and willingness to participate in independent actions in implementing the principles of the Global Compact. The Global Agreement is based on the principles of human rights, labor, and the environment, which are listed in the following documents:

- *The Universal Declaration of Human Rights;*
- *The International Labour Organization Declaration on Fundamental Principles and Rights at Work;*
- *The Rio de Janeiro Declaration on Environment and Development;*
- *The United Nations Convention Against Corruption.*

Certain forms of social responsibility are relevant to certain business areas: it depends on the nature of the activity and contact with the public. This contact is very important for organizations that seek not only a positive image but also a consistency in the organization. In order to ensure that managers do not violate the principles of sustainable leadership, first and foremost, they should focus on the corporate social responsibility of companies. Non-compliance with CSR in the future may lead to the loss of competitiveness. A socially responsible company should be responsible for every activity that has an effect on people, their communities, and the environment. It is safe to say that corporate social responsibility is an inevitable necessity of today.

Socially responsible companies need to make a profit, but they must do it ethically and honestly, by paying attention to the needs of society (Derry and Green 1989). CSR is measured and institutionalized on the basis of four key parameters:

1. markets;
2. employment (internal CSR);
3. public; and
4. environmental protection (external CSR).

Based on these parameters, CSR strategies are being developed, specifying and adapting corporate management to specific conditions.

Socially responsible behavior of the organization is created by selecting values (principles, norms, rules) according to their ability to promote functionality, economic efficiency, and business competitiveness; at the same time, such optimization of activities must be combined with human rights, dignity, the unconditional value of life, etc. Thus, there is a need for business ethics when pragmatism and morality, economy, and ethics are combined. We see that social responsibility is inherent in the concept of *values*; the same can be said about the organizational culture (Doane 2005). Applying responsible business practices may help the company to create a competitive advantage, to provide new opportunities, and to reduce operating costs (Ali, Rehman, Ali, Yousaf, and Zia 2010). It is stated that it is costly to work responsibly, but the application of social responsibility principles to a company is not necessarily more expensive; on the contrary, professional responsible business practices can improve the company's profitability. CSR also improves the image of the company. Publicly available information about corporate social responsibility helps to build trust and to create an attractive and trustworthy employer image. A socially responsible company can expect to have the best employees, not only to attract but also to retain staff. This reveals clear links among sustainable or socially responsible companies, high organizational culture, and sustainable leadership.

State priority concerning CSR is to promote corporate social responsibility while working with economic, social, and international partners. Since 2011, in EU strategic documents, corporate social responsibility is not considered just as a voluntary business initiative, because the business alone is not able to ensure the growth of competitiveness and the benefits of the CSR to society. In October 22, 2014, the European Parliament (EP) and European Commission adopted a directive on disclosed non-financial information of corporates (PE-CONS 47/1/14), which provides that in their annual accounts companies should disclose more extensive information on the sustainability of their activities: the key social and environmental factors affecting their activities, their fair and comprehensive policies, and an overview of results and risks, as this should boost investor and consumer confidence. In 2016, Lithuania has enacted legislation implementing the new EU directive on mandatory disclosure of non-financial and diversity information by large companies (public interest entities) and groups of companies. Since 2017, Lithuanian companies who have over 500 employees will be required by law to provide non-financial information covering environmental, social responsibility, human rights, and anti-corruption issues. It will also be required to describe the diversity policy applied to the administrative, management, and monitoring bodies of companies, revealing such aspects as age, gender, education, professional experience, etc. The requirements of the directive cover key areas of social responsibility and refer to the core principles, standards, or reporting guidelines of social responsibility, (e.g., ISO26000, EMAS; UN Global Compact, Global Reporting Guidelines [GRI]). The new directive seeks to consolidate and describe in detail the ways in which large companies should disclose this information in order for it to be comprehensive, objective, and not limited to a descriptive report of the enterprise's activities, but to provide numbers that are measurable by performance indicators in a quantitative manner, in order to allow comparisons between different companies and the economy branches or countries. However, the implementation of the directive will cause problems, as all these recommended systems are very different and reflect different business achievements, despite the fact that all of them are designed to measure a company's progress in the field of social responsibility. It is obvious that companies that will be covered by the directive will incur additional costs because additional competencies are needed in order to disclose qualified non-financial information.

It is therefore very important to ensure that the government takes the necessary statutory acts and other related decisions in order to implement the directive responsibly. By making the business burden more important by non-financial reports, entrepreneurs will be forced to shift the burden of increased costs onto the shoulders of consumers, and the deprived Lithuanian population will hardly be able to voluntarily support socially responsible entrepreneurs with their consumption decisions. It is necessary to recognize that, in Lithuania, when choosing a product or service, such criteria as the goods and the price and quality of the service still prevail, rather than the reputation or benefit to society of socially responsible companies. Therefore, additional measures are needed to support socially responsible businesses and to increase their competitiveness, such as reducing the tax burden on socially responsible companies and other forms of subsidisation, including the CSR criterion in public procurement, etc.

Meanwhile, the promotion of CSR initiatives in Lithuania focuses primarily on the development of social responsibility coordination platforms, educational tools, the development of interactive publicity tools, the dissemination of advanced information, the development of systems for measuring social responsibility results, etc. Priorities for such measures are set out in the Corporate Social Responsibility 2015–2020 Action Plan. This is reflected in its main objectives: to set up a platform for the coordination of corporate social responsibility between public authorities and social partners and to instruct it to monitor the actions of this action plan; to raise awareness of corporate social responsibility among companies and society, with a particular focus on companies operating in counties; to create conditions for continuous cooperation between municipalities, non-governmental organizations, and businesses; to ensure that public authorities recognize socially responsible companies; to ensure measurement of corporate social responsibility results according to objective indicators and improvement of availability of corporate social responsibility data. The Corporate Social Responsibility 2015–2020 Action Plan is financed from EU structural funds and state budget funds.

As currently no one doubts that socially responsible business meets public welfare expectations and benefits the country's social and economic development and the successful achievement of sustainable development objectives, the government must financially promote and support socially responsible business, while in Lithuania the promotion of CSR development is understood only as implementation of information dissemination and publicity measures. However, in a country like Lithuania, where the purchasing power of the population is very limited, this is not enough. Obviously, there is a need for additional incentives for socially responsible business because the benefits to society or business responsibility are not a priority criterion for consumers in choosing a product or a service.

Therefore, a business can also make a significant contribution to the creation of an equal, healthy, and vibrant society. The concept of corporate social responsibility allows us to justify the importance of sustainable business development in achieving the goals of sustainable development of the country; thus, the following part of the book reviews the most important principles of the concept of corporate social responsibility, focusing on the social responsibility of socially responsible companies towards their employees.

# 2 Corporate Social Responsibility and Social Commitments to Employees

## 2.1 CONCEPT AND LEVELS OF CORPORATE SOCIAL RESPONSIBILITY

Social responsibility and ethics are one of the aspects of the connection between business and society (Boulouta and Pitelis 2014). The concept of corporate social responsibility is based on the fact that management is viewed not only as an economic institute, focused solely on profit-making, but also as a part of society; therefore, it is responsible to society. It is important to assess the organization's relationship with the owners, investors, employees, customers, environment, and the whole community. Corporate social responsibility forces business representatives to be responsible for their actions. Society expects social responsibility from the business; thus, many organizations consider social objectives in their activities.

Social responsibility is a responsibility based on the company's datum points, including the input to social welfare and responsible attitude, taking into account the surrounding environment and the interests of the parties involved in the dialogue. In addition, it also serves as a competitive advantage influencing the reputation and success of the company. An essential condition of social responsibility is that general management must comply with valid principles and their implementation (Basu and Palazzo 2008).

Sustainable business is primarily a socially responsible business, which is scientifically formed by balancing social humanistic (ethical) values and economic achievements. As already mentioned, the base of corporate social responsibility consists of three underlying elements, which are *economic, environmental*, and *social* (Table 2.1).

*Economic responsibility* implies the consideration of shareholders' profitability objectives and their input in implementing economic welfare in society. *Environmental responsibility* signifies the conservation of the environment and natural resources. *Social responsibility* indicates taking care of employees, showing how open the company is, revealing that it is in good practice relations with stakeholders interested in dialogue, as well as displaying how responsibly and ethically it takes into account the opinions of stakeholders in the dialogue. A successfully operating organization adapts all these social responsibility aspects so that they can

**TABLE 2.1**
**Elements of Corporate Social Responsibility**

| Economic responsibility | Environmental responsibility | Social responsibility |
|---|---|---|
| Cost-effective activity—profitability | Knowledge of the environmental impact of business activities (raw material usage, environmental pollution, waste) | Taking care of the employees' welfare, development, and motivation |
| Competitive goods and services | Constant improvement of the activity | The maintenance of open communication in the dialogue between stakeholders |
| Financial risk management | Environmental legislation knowledge and observance | Promotion of good practice and cooperation |
| | The identification of the necessary changes and their implementation | Looking after the needs of society, business partners, suppliers, and customers |

*Source:* created by authors.

function and at the same time maintain balance. All of the above-mentioned elements are significant for sustainable business development and shall be equally and continuously developed at the same time (Bird, Hall, Momente, and Reggiani 2007).

In the modern world, CSR is becoming a new form of regulation (effective self-regulation), both at the macro and at the micro levels. This is determined by the changes in values that have taken place in the whole civilized world over the past 50 years (Boehe and Cruz 2010):

- As the societies are becoming more democratic and human rights realization is expanded into all spheres of activities, the dynamics of work relations is being monitored: from resigned relationships to associative and mutually respectful communication; from blind obedience to the authority to a perceived commitment to commonly accepted values.
- From the attitude and the principle of activities "only the strongest survive," "the smarter the better," "everyone against everyone," "man is a wolf to another man" to participation, care, cooperation, subsidiarity and solidarity principles.
- From the Marxist "class struggle" paradigm to the open—contractual—horizontal relations. From revolution and barricades to the reconciliation of interests and partnerships, i.e., from the hiding of interests (and "going over someone's head") to their transparency, legitimacy, and the adjustment and fostering of legitimate interests of individuals.
- From rapid profits to long-term, sustainable welfare.

That is why corporate social responsibility is considered as a *fundamental condition for sustainable development.*

In the organization, the realization system of sustainable business development, thus also the CSR, can be presented in a theoretical model combining factors that promote corporate social responsibility and the consequences of CSR activities on the overall performance of the company (Rehman, Rodríguez-Serrano, and Lambkin 2017). Such a model combines five essential components:

1. corporate and business unit strategies;
2. actions of sustainable development;
3. sustainable development actions linked by sustainable development activities;
4. the reaction of stakeholders;
5. performance of the company.

These components interact in the following way: sustainable development factors (defined in the strategies) promote actions that are combined in the activity of sustainable development into which various stakeholders react and, in this way, affect the financial result.

In order to create a sustainable business model and to adjust it to the company, it is necessary to identify the key stakeholders (Bhattacharya, Sen, and Korschun 2008; Collier and Esteban 2007). According to M. Freeman, the author of the stakeholder theory, interest groups are groups or individuals that can influence the organization or are influenced by organizational actions and pursuit of the objectives themselves. The activities of each company have an impact on society, but, first and foremost, on the interest groups associated with it, some of which are closer, while others are indirectly affected. Each stakeholder group has its own interests. An organization is considered responsible if it respects the interests of all stakeholders (Figure 2.1). The CSR concept must be developed and promoted in the interests of both the business organization and other stakeholders (Freeman, Harrison, and Wicks 2008; Godfrey, Merrill, and Hansen 2009).

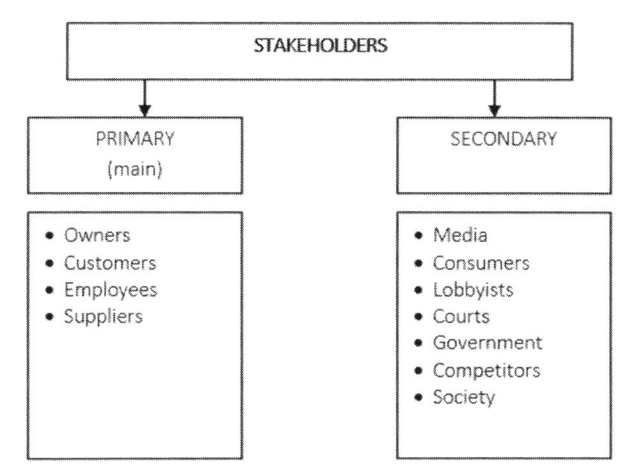


**FIGURE 2.1** Stakeholder distribution. (Source: created by authors.)

The CSR model presented by the European Commission (Figure 2.2) allocated internal and external dimensions of corporate social responsibility:

1. Internal dimensions:
    - *Human resource management* includes recruitment of qualified labor, lifelong learning opportunities for employees, provision of balance time between work, family, and leisure, and assurance of fair pay and career opportunities independent of gender or age.
    - *Health and safety at work.* Ensuring security and minimum risks at work has become an important criterion in assessing the performance of companies.
    - *Adaptability to change* means that employees must be assisted in the restructuring, mergers, divisions, and similar processes of the company.
    - *Management of the impact on the environment and natural resources.* The reduction of environmental pollution and effective use of resources must be organized in a way that improves profitability and competitiveness by ensuring business success without harming the environment.
2. External dimensions:
    - *Local communities.* Businesses help local communities in creating new jobs, paying wages to employees, and contributing to community welfare by paying taxes. Companies are directly dependent on

**FIGURE 2.2** Dimensions of corporate social responsibility. (Source: created by authors.)

communities that work for the sake of health, stability, and the common good. Therefore, mutual relationships and benefits must be properly balanced.
- *Human rights.* Pressure on the company to apply CSR principles in the field of human rights can be felt from public or non-governmental organizations, consumers, or other stakeholders.
- *Business partners, suppliers, consumers.* Corporate social responsibility objectives are influenced by the partners, suppliers, and consumers of the company. Consumers, as one of the key stakeholder groups, must be given the highest quality, security, and reliability.
- *Global environmental protection.* Since many businesses and organizations operate on a global scale, their business impact is rarely limited to one country. Thus, the principles of social responsibility must not be applied only in one country or in Europe alone; they must be implemented globally, and internationally.

In view of the indicated dimensions, organizations should identify the social consequences of their activities together with their impact on the environment and evaluate them properly. This, in turn, would allow organizations to improve their social and environmental management activities.

CSR is best defined as a process of a broad spectrum, covering the entire product and/or service production and delivery cycle together with environmental, social, financial, and ethical actions related to it and provisions which are voluntarily pursued and fulfilled by the companies participating in this cycle (Du, Bhattacharya, and Sen 2011; Dahlsrud 2008). Thus, an essential issue is the benefits of CSR for the company and the development of the country's economy and society.

One can distinguish the following key statements, supporting the benefits of socially responsible business (Birth, Illia, Lurati, and Zamparini 2008):

1. *Creation of favorable business long-term perspectives.* Social actions of companies improve the lives of local communities and reduce the need for state regulation. In a socially successful society, there are better conditions for the business. In addition, although short-term costs related to social activities may be high, they can increase profits in the long run, as an attractive brand image is created in the sight of the local community, consumers, and suppliers.
2. *Change in society's needs and expectations.* In order to reduce the gap between new public expectations and the real response of companies, business has become increasingly more involved in social problem-solving, as after assessing the benefits and disadvantages of globalization together with the business role in this process, it became clear that many social problems cannot be solved without the support of business.
3. *Resource availability and their allocation to address social problems.* Business has considerable human and financial resources at its disposal and, therefore, a part of them should be transferred to meet social needs (e.g., the population health, their professional background, motivation to work qualitatively, etc.).

4. *Moral commitment to implement socially responsible activities.* The company is a member of society; therefore, organizations, as individuals, have to act in a socially responsible way in order to strengthen the foundation of public morality.
5. *Consolidation of human resources and intellectual capital.* CSR is an important factor in boosting employee motivation as well as attracting and retaining highly qualified personnel.
6. *Assurance of reputation and safety.* Reputation is based on the following intangible things: trust, quality, consistency, relationships and their transparency, as well as defined things: investment in human capital.

CSR is not a substitute for public policy, but it can effectively help to achieve a number of public policy objectives, such as:

- more integrated labor markets and a higher degree of social integration, since companies actively recruit people from disadvantaged groups;
- investment in skills development, lifelong learning, and employability opportunities, which are necessary to ensure competitiveness in the global knowledge economy and to address the issue of the ageing population in Europe;
- public health improvement, driven by voluntary business activities;
- high level of innovations that address societal problems through enhanced cooperation with external stakeholders and a working environment that is more favorable to innovations;
- expedient use of natural resources and reduction of pollution, in particular through investments in eco-innovation and implementation of voluntary environmental management systems;
- better image of business and entrepreneurship in society, assisting in the development of positive attitudes towards entrepreneurship;
- assurance of human rights, environmental protection, and core labor standards, especially in developing countries;
- poverty reduction and progress towards the Millennium Development Goals.

The following key elements can be distinguished that hinder the development of social responsibility:

- *Denial of the principle of profit maximization.* The allocation of a part of the profit for social purposes would reduce the effectiveness of the principle of profit maximization. The company behaves in a socially responsible way, focusing only on economic interests, and social issues are a state prerogative.
- *Social inclusion costs.* Funds allocated for social needs mean additional costs for the company. These costs, transferred to the consumer's shoulders, would increase prices for the provided services and goods.

- *Insufficient level of public responsibility.* Since society does not elect the managers of companies, they are not directly responsible to the general public (hence, they cannot be held responsible).
- *The lack of ability to solve social problems.* The company's staff does not have experience in solving social problems. Public development should be carried out by professionals working in the relevant state institutions and charities.

One can distinguish four arguments, proving why companies should apply the principles of social responsibility:

1. Social problems are caused by organizations, so they have to be responsible for solving these problems and trying to prevent new ones.
2. Organizations have more resources and powers that should be used to support social welfare.
3. Organizations cover many social aspects (recruitment, product development, etc.) in their activities that force them to act responsibly.
4. Organizations must meet not only their own expectations but also the expectations of all the stakeholder groups concerned with the organization's activities.

Summarizing the arguments for and against socially responsible business, one can assess the benefits and costs of the organization when implementing the CSR principles (Table 2.2).

Having evaluated all the arguments, the following key aspects that motivate companies to deploy CSR can be distinguished:

- opportunity to better meet the needs of your consumers (customers, patients, etc.);
- creation of a good brand name and the reputation of a reliable company, thanks to which it is possible to:
  - raise the loyalty level and strengthen links with the local community and government;
  - increase the sales volume of your goods or services;
  - increase employee motivation and labor efficiency in order to:
    - cultivate human resources;
    - reduce costs;
    - increase competitiveness.

By pursuing socially responsible activities, organizations justify public expectations, gain greater competitive advantage, and can ensure higher profitability (Blomback and Wigren 2009).

Thus, it can be concluded that socially responsible behavior is becoming cost-effective for companies, as socially responsible business gives companies a significant competitive advantage, ensuring a good reputation of the company, while the social acceptability of the economic performance becomes an essential condition

**TABLE 2.2**
**Benefits and Expenditure of Corporate Social Responsibility**

| Stakeholders | Benefits | Expenditure |
|---|---|---|
| Directors | Greater independence, autonomy | More meetings and instructions |
| Shareholders | Increased investment from ethically based pension funds | Increased reporting costs, more transparency, openness |
| Managers | Better human resource (HR) policy measures increase employee motivation<br>Better knowledge of ethical issues enhances confidence in employees | More intense ethical learning<br>Participation in focus groups and reporting |
| Employees | Better HR policies allow increasing motivation<br>The correct and ethical behavior of managers makes it possible to increase productivity<br>Fewer work-related disputes and strikes<br>Better working conditions<br>The choice of CSR allows the company to attract highly skilled employees and young people more easily<br>Lower costs for the search of the employees and recruitment | Increased focus and expenditure on ethical learning<br>More communication among the employees in the company<br>More effort is needed for labor relations<br>It is necessary to follow policies, complying with human rights |
| Customers | The shift to ethical consumption<br>Fewer disputes<br>Advertising can be based on CSR<br>Protected reputation<br>Recognition of brand quality | Commodity prices may increase slightly in the short term |
| Subcontractors/ Suppliers | Better quality contribution<br>Increased confidence<br>More flexible relationships | The cost of the provided resources may increase in the short term |
| Community | Greater willingness to take in new investments | Continuous communication with the community is needed, which requires additional financial resources and time |
| Government | Increased confidence in the company<br>Fewer legal disputes, no new harmful laws are created<br>More favorable trade regime<br>More favorable acceptance of development and reduction<br>More effective cooperation with social policy institutions | Costs for new regulations may increase<br>Government expenditure on social and labor market policy measures may increase |
| Environment | Fewer legal disputes and fights<br>Improved public image<br>Improved working environment<br>Improved health of employees | Investments in environmental damage control<br>Investments in environmental technologies |

*Source:* created by authors.

for the long-term stability and success of the company (Boehe, Cruz and Ogasavara 2010). However, as mentioned earlier, society should have achieved a certain level of development, where the priorities when selecting goods and services are not only the price and quality ratio but also a sufficient amount of attention given to the reputation of the company, benefits to society, and the achievement of sustainable development objectives.

The CSR implementation tools play an important role; therefore, they are discussed in more detail in the next section.

## 2.2 TOOLS OF CORPORATE SOCIAL RESPONSIBILITY IMPLEMENTATION

CSR is a process of a broad spectrum covering the entire product/service production/development as well as associated environmental, social, financial, and ethical aspects. Social responsibility can be considered as an application of ethics, sustainability, and responsibility principles in daily corporate activities:

- *Ethics* is vision, objectives, values, organizational structure, behavior, and culture.
  - *Sustainability* is business processes, supply chain, production, product design, and distribution.
  - *Responsibility* is a relationship with various stakeholders: government, employees, customers, suppliers, and society.

As mentioned in the previous chapter, the basis of social responsibility consists of three elements: economic, environmental, and social (Saeidi, Sofian, Saeidi, Saeidi, and Saaeidi 2015). *Economical responsibility* is profitable companies supplying competitive goods and services throughout the changing global economic conditions, as well as perfectly managing financial risks. Corporate value is a key indicator of the economic responsibility of companies. *Corporate environmental responsibility* is the knowledge of the environmental impact of company activities, constant improvement of activities, implementation of innovative technologies that are environmentally friendly, and compliance with environmental legislation. *Social responsibility* is the care for the welfare, improvement, and motivation of the company's employees, care for the needs of society and customers, and maintenance of open communication with stakeholders. It can be summarized that with all three of its elements CSR primarily makes an impact in these key areas: the workplace, the community, the natural environment, and the market (Waddock and Googins 2011).

*Social responsibility in the workplace:* safe and healthy workplace for employees as well as the promotion of awareness in this area; respect for and protection of human rights in the workplace, equal working conditions for representatives of various social groups; opportunities for employees to learn for life, improve themselves and help others to improve. Hence, the implementation of social responsibility in the workplace focuses on the creation of positions of employment without discrimination, that promote learning and professional attitudes allowing a balance of work-life and other aspects of life.

*Social responsibility in society and the community:* paying attention to the needs of the local community, finding a compromise between interests of the community and company; philanthropy and voluntary participation in activities or initiatives of the community and society; motivation of young people to participate in business and professional activities, knowledge transfer and the creation of opportunities for practice. Thus, companies can ensure safe and healthy working conditions, reduce the negative influence of services and products, and actively participate in public activities, as well as promote regional development, education, and healthcare in order to maximize the given amount of funds, time, and other resources.

*Social responsibility in the natural environment:* effective and responsible use of resources in the activities of the company; protection of the environment and revitalization of abandoned areas during the development of activities; development and production of ecological products (suitable for recycling, consuming fewer resources in the process of production, promoting economical use of energy and other resources).

*Social responsibility in the market:* CSR activities, related to the market and the economic impact on it, include such aspects as the impact of products and services, ethical trade, transparency of competition, and support for a fair globalization. Dutiful payments in accordance to incoming bills; socially responsible marketing that does not take advantage of weaknesses of different social groups; paying attention to specific needs of different groups (people with disabilities, youth, pregnant women, etc.).

Changes in the global environment inevitably lead to sustainable development and social responsibility; therefore, these management conceptions inevitably change, overlap, and add up to each other's issues. Recent research has shown an increased interest in CSR and the expression of sustainable development concepts in global quality management. It can be argued that the development of social responsibility is a change in the company's values by focusing not only on the needs of the individual but also on the needs of society.

By analyzing works of many authors, the importance of the manager's approach to social responsibility becomes clearly noticeable because socially responsible behavior mostly depends on moral and ethical provisions and values of people in charge; however, their attitudes are not enough. First of all, CSR must be the most crucial element of values of the employees of the company as well as an important factor of organizational culture (Saeed and Arshad 2012; Rosińska-Bukowska and Penc-Pietrzak 2015).

There is no doubt that the level of implementation of social responsibility initiatives in the organization is determined by the ethical attitudes of its managers, business culture, and philosophy, as well as the interests of the society that influence business orientation not only for the direct economic objectives of companies but also for social and environmental objectives, as well as efforts to increase business development possibilities (Figure 2.3).

CSR can be considered a commitment that suits the value system of the company, its operating policies, and moral principles of managers, including socially responsible and ethical management of the company, effective use of resources,

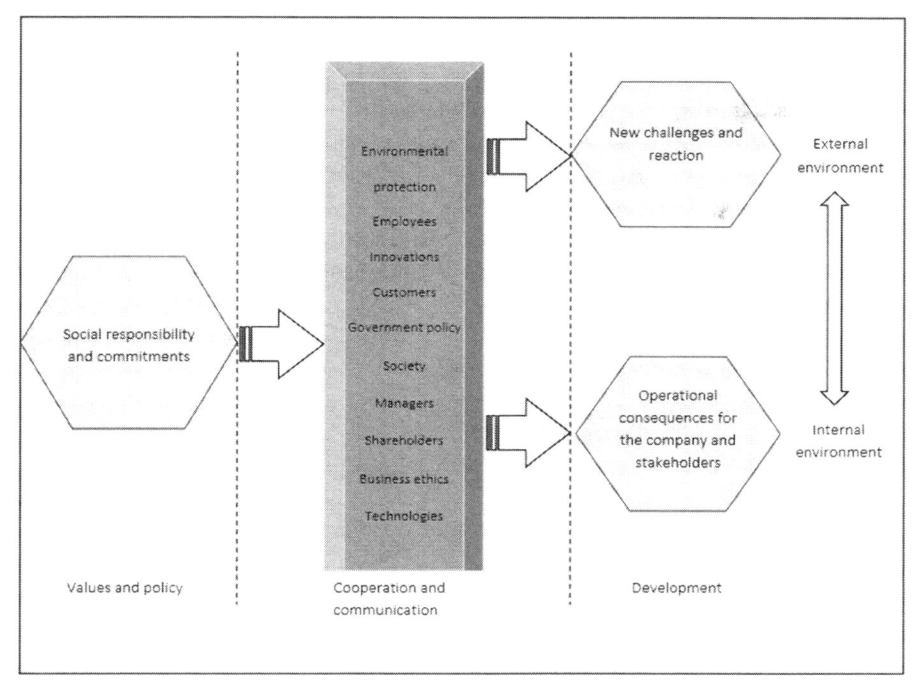


**FIGURE 2.3** Expression of social responsibility principles in the activities of the company. (Source: created by authors.)

advanced technologies, respect for the environmental requirements, interests of employees, customers, and society, as well as high level consumption culture and socially responsible investment (Boyd, Bergh, and Ketchen 2010). Balancing all these expectations of stakeholders allows companies to react more effectively and rapidly to external environmental changes as well as to reach the objectives of the company. CSR manifests itself as close cooperation and communication between the internal environment of the company and the external environment (society). Companies must strive to exploit all realization opportunities of CSR concepts and must understand that the implementation of CSR principles in activities of the company benefits the final results (Kim, Lee, Lee, and Kim 2010).

The Global Compact, initiated by the UN, is the key tool for developing corporate social responsibility. As mentioned before, the purpose of this agreement is to motivate businesses to conduct their business responsibly, not to harm the environment, community, or other businesses, as well as in the cooperation with the UN, governments, and non-governmental sector to contribute to solving specific social and environmental problems, to societal development and in such a way to contribute to the socially inclusive economic development. The initiative enables companies and organizations to share knowledge, experience, and innovations on how to improve their business strategy and corporate image, and to improve risk management strategy. The Global Compact is not designed to oversee, monitor, or measure the behavior and actions of companies. The Global Compact is based on public responsibility, transparency, and interest, as well as the willingness of companies to participate in independent actions by implementing principles of the Global Compact (Table 2.3).

In order to become a member of the Global Compact, an organization has to anticipate its business development in such a way that principles of the Global Compact would become a part of the strategy, culture, and daily work of this company; in annual reports, activities that support the Global Compact and its ten principles have to be described/indicated. By participating in the Global Compact, the organization:

- shows its leadership as it supports global principles and responsible business ideas;
- contributes to the development of practical solutions to globalization and sustainable development problems;
- controls risks by preparing preliminary plans of actions for the potential critical situation;
- improves management of organization/brand, morale and productivity of employees, and the effectiveness of the performance;
- has an opportunity to share its experience and knowledge;
- has access to the UN information and knowledge about development processes and their practical aspirations in the world.

**TABLE 2.3**
**Principles of the UN Global Compact**

| Area | Principles |
|---|---|
| *Human rights* | Principle 1: businesses should support and respect the protection of internationally proclaimed human rights.<br>Principle 2: ensure that they are not complicit in human rights abuses. |
| *Labor* | Principle 3: businesses should uphold the freedom of association and the effective recognition of the right to collective bargaining.<br>Principle 4: the elimination of all forms of forced and compulsory labor.<br>Principle 5: the effective abolition of child labor.<br>Principle 6: the elimination of discrimination in respect of employment and occupation. |
| *Environment* | Principle 7: business should support a precautionary approach to environmental challenges.<br>Principle 8: undertake initiatives to promote greater environmental responsibility.<br>Principle 9: encourage the development and diffusion of environmentally friendly technologies. |
| *Anti-corruption* | Principle 10: business should work against corruption in all its forms, including extortion and bribery. |

*Source:* created by authors.

At the national level, the Global Compact is based on the creation of a national network. This network usually connects businesses that consider themselves socially responsible. Representatives of the companies participating in this network regularly meet and discuss relevant issues (internal aspect: an improvement of an activity or business of a company); if companies decide to contribute to solving specific problems in society, they often carry out large-scale projects aiming at long-term outcomes or impacts in a particular area (e.g., environment) and not only to short-term advertisements or impacts that come from philanthropic activities of different companies.

Participation in the Global Compact can be beneficial for business because:

- *Being a member of the Global Compact means that company is globally recognized as socially responsible.* This is important for the company's image and increases its competitiveness (e.g., for Lithuanian companies to establish themselves in the EU market, they are also assessed according to whether they conduct their business responsibly, i.e., do not violate human/employee rights, avoid polluting nature, etc.).
- *The company can improve its image because of the Global Compact.* Examples of some companies show that, because of the image of a socially responsible company, it can become very successful in business.
- *The Global Compact gives the opportunity to exchange experiences on issues/problems that are relevant to companies.* The Global Compact Office organizes annual worldwide meetings with business representatives.
- *A company implementing the principles of the Global Compact carries out the risk management strategy.* A company that does not implement risk management in its business strategy cannot expect success from a long-term perspective.

Thus, participation in the Global Compact is beneficial to the competitiveness of the company and its establishment in international markets; also, it provides the organization with an opportunity to participate in political dialogues, training, local networks, and partnership projects based on the implementation of principles and the UN development objectives.

Systematic management requires the company to accurately define its values, related management principles, and actions. In order to enhance its performance, the company has to choose the appropriate indicators that can also be used to present results. These operations are facilitated by the implementation of various management systems (quality, environmental, social management, etc.) into a single integrated management system of actions.

On November 1, 2010, ISO issued ISO 26000 Social Responsibility Guidelines/Recommendations standard. ISO 26000 supplements already existing initiatives of social responsibility and expands the understanding and application of social responsibility. Basically, ISO 26000 summarizes the whole modern definition of social responsibility, covering many issues related to it. Moreover, ISO 26000 is a standard of guidelines, not requirements, and is not intended for certification.

The main advantage of ISO 26000 guidelines for companies is a clear model and set of recommendations based on which the company can better evaluate its activities in the area of social responsibility:

- whether the principles of social responsibility are implemented in all areas (processes) of the company;
- whether the existing maturity of corporate social responsibility corresponds to the examples of leading practice and what actions should be taken by the company in order to ensure the development of activities and improvement of the company in the social responsibility area.

The Social Responsibility Guidelines standard aims primarily at creating added value by structuring and presenting universally accepted principles in the area of corporate social responsibility (Brammer, Jackson, and Matten 2012). Guidelines are for presenting universally accepted principles but not for replacing already existing international agreements such as the Universal Declaration of Human Rights or the International Labour Organization's Declaration on Fundamental Principles and Rights at Work. Presented social responsibility principles are divided into general, main, and operational principles.

*General principles.* These principles include respect for internationally acknowledged conventions and declarations, compliance with legal requirements, and a recognition that stakeholders have the right to be heard and organizations should take their opinions into account.

*Main principles.* These principles are related to the organization's impact which the company is held responsible for. These principles consist of such areas as environmental protection, human rights, recruitment and work practices, management, transparent activities, and incorporation of community and consumers.

*Operational principles.* These principles set guidelines for the activities of organizations. By taking into account the interests of many stakeholders, their transparency and approach to life-cycle, these principles include limits of accountability, integration, and significance.

ISO 26000 presents companies with measures and tools for self-evaluation and further CSR development in their own activities. Other standards provide management tools to ensure compliance of companies in each of the areas of social responsibility: ISO 9001—Quality Management Systems (Requirements); SA8000—Social Accountability 8000; OHSAS 18001—Occupational Health and Safety Management System; ISO 14001—Environmental Management Systems; ISO 50001—Energy Management.

The management system of social responsibility might be created in order to manage social responsibility issues (Dutot, Lacalle Galvez, and Versailles 2016). During this process, the company can use quality management standards (ISO 9000), environmental management standards (ISO 14000), energy management standards (ISO 50001), social management standards (SA 8000), etc. The best solution is to combine systems of different standards into a single operational system. Environmental and social security management is related to quality management principles. Therefore, environmental and social security systems can be successfully integrated with standardized quality management systems.

Social and environmental management systems and standards are the best examples of a social transaction on the micro level (Lee 2008). The environmental management system is one of the elements forming the overall organization management system, which also includes the organizational structure and the method of determining responsibility, as well as the order, process, and resources of environmental policy-making and implementation (Lin, Yang, and Liou 2009; McWilliams, Siegel, and Wright 2006). Based on the environmental management system, the established priority environmental problems are being managed. By introducing these systems, organizations are encouraged to use their environmental potential as effectively as possible, to organize preventive activities, and to ensure continuous progress by increasing the effectiveness of environmental protection.

Recently, several standardized environmental management systems have widely spread throughout the world with environmental, energy, and social SA 8000 management systems gaining the widest recognition.

In the whole world acknowledged international environmental management systems standard ISO 14001 belongs to the general group. This standard is applicable to any organization operating in any sector of industry that seeks to reduce the environmental impact of its activities. This standard requires organizations to identify all environmental impacts and related aspects and then take action to improve processes in priority areas that have significant aspects.

The following benefits of the environmental management standard ISO 14001 can be distinguished:

- high-quality environmental management;
- greater reliability;
- higher awareness of employees;
- better image of the organization in society;
- new business opportunities;
- increased customer confidence;
- improved relationships with customers, the local community, and control authorities.

By implementing ISO 14001 principles economic efficacy increases, environment-friendly conditions are formed, and a new competitive advantage is created in international markets. Organizations, voluntarily and at their own expense, decide to implement this modern management system, as it strengthens their export position in the competitive international market.

ISO 50001 is a standard for increasing energy efficiency. Every year, energy resources are becoming more expensive and organizations' energy costs make up an increasing share of the expenditure. By reducing energy consumption, operating costs and greenhouse gas emissions are also being reduced. Efforts to reduce energy consumption also demonstrate that the organization is concerned about natural resources and protection of the environment. Thus, ISO 50001 is an Energy Management Standard designed primarily to control energy costs and reduce greenhouse gas emissions. The energy management standard provides structured and comprehensive methods for improving energy efficiency, i.e., this standard provides the requirements for the design, implementation, maintenance, and improvement of the management system to continuously observe and reduce energy consumption.

ISO 50001 operating principles:

- In order to increase energy efficiency, energy management policy of the organization is defined.
- Energy-saving objectives and tools to achieve them are set.
- In order to make appropriate decisions, energy consumption data are analyzed.
- Results are measured.
- The energy management system is evaluated on how it works.
- The energy management system is constantly improved.

The energy management standard ISO 50001 is addressed to all organizations; however, organizations that consume a lot of energy would benefit the most from it: industrial companies, energy producers, water suppliers, waste processors, real estate managers, hospitals, and shopping malls.

BS OHSAS 18001 is a standard issued by the British Standards Institution and it belongs to the Occupational Health and Safety Management Systems family of standards. The introduction of OHSAS demonstrates a responsible organization's approach to occupational safety. In many countries, the Occupational Health and Safety Management System is mandatory. The organization has to evaluate the safety and health risks related to its activities, understand how to control these risks, and set clear objectives in order to improve the continuous efficiency of the system. This standard helps organizations to control threats in usual operational processes and avoid critical situations. Moreover, this standard covers three important components: hazard identification, risk assessment, and risk control.

*Hazard identification.* Analysis of conditions and situations in which employees might get injured or fall ill.

*Risk assessment.* Evaluation of risks arising from potential hazards and the analysis of impacts from combinations of hazards.

*Risk control.* Control of potential risks in order to keep them at an acceptable level.

First of all, the organization benefits from OHSAS 18001, as it ensures compliance with legal requirements. The organization in which the Occupational Health and Safety Management System is implemented fully complies with the legal requirements for occupational safety. In addition, a structured approach to the management of occupational safety and health risks is created and risks are continuously measured and controlled. The standard helps to improve the organizational culture because, with the introduction of hazard risk system, employees themselves are more concerned with the prevention of hazards and risks. In addition, after implementing this standard, the appeal of an employer increases, since an organization that cares about health and safety of employees attracts potential employees

(Frostenson, Helin, and Sandström 2011). OHSAS 18001 is easily compatible with other standards such as the quality management standard ISO 9001 and the environmental management standard ISO 14001.

Standard SA 8000 is a document issued by the US Council on Economic Priorities Accreditation Agency (CEPAA). This standard lays out requirements for the ethical practice of employment based on 11 international occupational conventions of organizations, the Universal Declaration of Human Rights, and the United Nations Convention on the Rights of the Child. This standard was issued in 1997. The emergence of this standard was mainly influenced by the opinion of consumers. Consumers from developed countries show more and more concern about how the products they use have been produced. Companies started to issue their own codes of conduct in which the main humanized workplace guidelines were set and all suppliers and subcontractors had to comply with it. The main principles are to ensure proper working conditions, quality of products, management system, and environmental aspects. The Council on Economic Priorities issued SA 8000 standard because of the increase in the number of organizations with their own codes of conduct. SA 8000 is a social responsibility standard for companies of all sizes that want to take into account the social and ethical aspects of their business. The social responsibility system proves to customers that the organization is taking action to protect the rights of employees and ensure the ethical production of goods. The SA 8000 standard ensures that a company with this certificate has acceptable social working conditions. This standard is designed to evaluate and certify systems of organizations that ensure acceptable working conditions. Compliance with the standards is controlled by external certification companies according to clear, publicly known, and easily verifiable rules and laws. This certification informs consumers that the goods are produced in compliance with international occupational health and safety standards.

SA 8000 is a social standard. It defines the main requirements for the certification process and sets the required minimal social standards. In addition, the standard includes implementation of a management system in an organization as a tool of self-regulation. One could distinguish such advantages of SA 8000:

- The initiative of the SA 8000 certification is based on the widely applied and universally accepted standard.
- Compliance with the standard is controlled according to known rules and laws, thus, it is easily verifiable.
- The realization of suitable management systems encourages personal initiative from the supplier, the continuous process of improvement, and, at the same time, the long-term implementation of minimal social standards around the whole world.
- The certificate proves that working conditions are adequate in the company, and it gives an advantage over the company's competitors.
- The overall process is transparent, trustworthy, and reliable.

The introduction of minimal social standards in the international world of work is a complex task and it cannot be done at the same time. It depends not only on the right conceptions but also on the dialogue related to this issue without any negative prejudices among all groups of stakeholders such as manufacturers, global trading companies, trade unions, governmental organizations, and end users. Such dialogue gives the opportunity to learn from each other and to gradually improve occupational conditions around the whole world.

Main provisions declared in SA 8000:

1. Child labor. No employees under the age of 15 years old, or 14 years old in some developing countries. Absolute elimination of child labor.
2. Forced labor. No forced labor, including work in prisons or repayment of debts by labor.
3. Health and safety. To provide a safe and healthy working environment, ensure prevention from injuries in the workplace, and regular health and safety training and examination. A system ensuring the protection of health. Access to drinking water and bathroom cabins in the workplace.
4. Freedom of trade union activities and the right to collective bargaining. To respect the right to be in a trade union and to participate in collective bargaining in order to protect personal rights.
5. Discrimination. No racial, religious, ethnic, gender, sexual orientation, political, or age discrimination. No sexual harassment.
6. Disciplinary practices. No labor punishments, physical and psychological pressure, or verbal abuse.
7. Working hours. Depending on the law, but not more than 48 working hours per week with at least one free day. No more than 12 hours of overtime with bonus remuneration.
8. Remuneration. Remuneration per week must comply with standards.
9. Management systems.

In order to get these certificates, companies must meet all of these requirements. The SA 8000 management system requirement includes:

- social responsibility and working conditions policies;
- periodic analysis of management;
- appropriate appointment of management representatives;
- planning and implementation of standard requirements;
- evaluation and appointment of suppliers (subcontractors) according to requirements of social responsibility;
- response to the concerns of employees or other stakeholders regarding social policy and execution of standard requirements as well as appropriate corrective actions;
- procedures for communication with external stakeholders;
- possibilities for stakeholders to carry out inspections;
- recording of data demonstrating compliance with the standard.

Companies are certified in accordance with the SA 8000 standard. However, this is not a widespread practice, as large companies typically publish their own codes of conduct and check themselves or their subcontractors for compliance with these codes by themselves or through independent organizations.

The Institute of Social and Ethical Accountability (SEAI) takes a very important role in the social audit practice. With the purpose of strengthening the corporate social responsibility and ethical behavior of business and non-profit organizations, this international organization was established in 1996. The Institute of Social and Ethical Accountability created a methodology for social accounting; moreover, in 1999, it issued AA1000—Accountability 1000—standards. The aim of AA1000 standards is to link the definition and consolidation of the company's values with the development objectives of the business. According to this approach, learning and development of the company, as well as the strengthening of the organization, are essential elements. As a process standard, AA1000 accurately defines mandatory processes required for the explanation of decisions that influenced the activities of companies without focusing on the operational level the organization should achieve.

In addition, the UN initiated the appearance of the Global Reporting Initiative (GRI) in 1997. In 2000, the first 50 reports were prepared based on this method. In 2010, 4,000 reports were produced and in 2013, 11,946 reports were prepared.

The GRI initiative encourages organizations to implement sustainable development principles and execute operational accountability. Sustainable development within the organization is implemented and accounted by these aspects:

1. Economic aspect: organization's economic impact on the stakeholders and the regional, national, and global economy;
2. Environmental aspect: organization's impact on the environment, used resources, external performance, compliance with environmental requirements, and expenditure related to them;
3. Social aspect: work practice in the organization, rights of employees, society, and accountability of products.

In addition, general information about the organization is employed: the strategy of sustainable development, a general description of the organization, significant aspects, stakeholder involvement, management of the organization, ethics, and integrity.

Data provided according to GRI allow us to measure the benefits generated by the application of sustainable development principles within the organization and create added value for the company. The method provided by GRI helps to assess long-term risks and becomes an increasingly important tool for the preparation of long-term corporate strategies. The 150 indicators reflect the key and globally recognized international standards of various areas. These standards contribute to more responsible business and activities of organizations. In 23 countries, governments and market authorities indicate or recommend the use of GRI guidelines in policies or regulatory framework of organizations. The implementation of an EU directive (which also refers to GRI) signals a positive influence of the methodology and a need to standardize accountability of CSR globally.

The G4 guidelines presented in 2013 are more convenient, simpler, and clearer. In May 2013, GRI became an official standard of the Global Compact of the UN and is recommended for all members to use as a tool for preparing progress reports. The methodology of GRI has been prepared on the basis of the principle of stakeholder inclusion. Acknowledged experts of respective areas from all over the world participate in its preparation and development. The methodology of GRI is based on the disclosure of the efficacy of a company in carrying out socially responsible activities. GRI indicators are reflected in the key globally acknowledged international standards that contribute to more responsible activities of companies. As already mentioned, acknowledged experts of respective areas from all over the world participate in the creation and improvement of the methodology of GRI indicators.

A report prepared in accordance with the GRI recommendations is not an objective but means for changing the organization. A unified methodology and a clear set of indicators help both the organization and the reader to evaluate the efficacy of the organization. The report integrates financial and non-financial disclosures. In addition, stakeholders are involved in the reporting process. In the report, significant aspects of the company are presented; thus, problematic areas can be identified and tools to solve those problems can be determined. The main principles of GRI content:

1. stakeholder inclusiveness;
2. sustainability context;
3. materiality;
4. completeness.

One of the key things when preparing a report is to identify significant operational aspects of the company. For example, aspects that reflect the major economic, environmental, and social impacts of the organization or have a significant impact on decisions and assessments of stakeholders.

Many researchers argue that leadership is one of the key principles allowing successful implementation of the concept of CSR and to ensure sustainable business development (Moura-Leite and Padgett 2011; Seitanidi and Crane 2009). The before mentioned *Natural Step* organization offered a strategic leadership model for sustainable development that can be implemented in every company. Of course, this model can also be used at the macro level for creating sustainable development strategies for the country or individual sectors, such as the energy sector, transportation sector, etc. Society keeps repeating the same mistakes. The problematic history of the industry, supply and consumption of materials, products, resources, and services highlight two important issues that have to be considered when planning:

- Often the impact is due to a complicated complex of interacting phenomena which could not have been predicted in advance. In the best case, some impacts can be attributed to certain activities or processes only after they have occurred. In addition, sometimes it is difficult to identify a specific impact of certain actions. Therefore, a long period of time passes between the identified impact and policy changes and this causes significant damage.

- This situation demonstrates a necessity to identify the principles of first priority based on which practical solutions should be assessed and strategies should be prepared to prevent damage. In order to make a decision, the preventive principle is the most important one, as strategies employed to fix damage are not sufficient enough to ensure sustainable development.

TNS, which is an international non-governmental organization, in cooperation with scholars, prepared a methodology for implementing sustainable development principles, which, based on the experience of human error, integrates the method of back casting from the main principles of sustainable development.

The scientific principles of the sustainable development model of TNS are not new. They are based on basic physical rules and the laws of thermodynamics. This model defines a sustainable society according to four conditions of the system, which describe a movement of material flows and the extent to which people's needs are met. Organizations, wanting to remain in the TNS model, are advised to implement this model in four stages. Organizations that operate according to these stages become strategic or pro-active and can avoid the withdrawal from the business by "hitting the walls of a funnel."

Back casting is a planning procedure that results in successful planning from future perspectives by asking: what should be done today in order to achieve successful results in the future (Holmberg and Robèrt 2000)? In other words, in order to achieve the objectives of sustainable development, current policy measures are planned. The basic principles of sustainable development provide the main principles according to which back casting is applied from sustainability achieved in the future. These principles include:

1. the necessity to rely on a scientifically based approach to the world;
2. the necessity to achieve sustainability;
3. the sufficiency to achieve sustainability;
4. general actions of the society that relate to sustainability;
5. specificity of action management and directive support for problem analysis;
6. no overlapping or denial of each other in order to ensure comprehensiveness and structural case analysis.

The methodology of back casting from basic principles was developed based on the methodology of creating scenarios. Back casting on the basis of scenarios is a planning methodology founded on the simplified image of success. This model of planning is based on a created vision that regulates the game and helps the player to deal with the problem of complexity; thus, it is reminiscent of the solving of a puzzle. This methodology is useful for creating a relatively static image of the future. This method of planning is useful for making emotionally charged decisions by asking decision-makers to create a specific image of the future that would allow supporting and weighing these decisions in a transparent manner. Such a method helps businesspersons, especially financial institutions that take moderate risks when making decisions. Even though back casting based on scenarios is a methodology that enables more strategically motivated, creative, and cooperative decisions, used to achieve a shared vision, it can pose some problems for sustainable development. A large group of people might have difficulties in reaching an agreement on a detailed vision of sustainable development because of different values, knowledge, etc. Technological development might change the conditions of planning by making scenarios impossible (Holmberg and Robèrt 2000). Detailed scenarios of sustainable development of a company might seem the opposite of sustainable.

Back casting, based on the principles of sustainable development, is similar to a chess game in which principles of success control decisions made in the game. The principle of checkmate in chess corresponds to the principles of sustainability in the implementation of sustainable development. It is a dynamic method of planning because every step evaluates the current game situation and at the same time optimizes winning opportunities. As in the case of chess, there is a large number of winning combinations and the chances of choosing a winning step increase significantly with every strategic step.

Back casting, based on the principles of sustainability, is a more convenient way to make decisions, because it is much easier to agree on basic principles of success and specific initial steps that are the basis of moving in the right direction and constantly assessing the success of the movement's path rather than agreeing on a detailed description of the distant future. Naturally, both principles can be combined (Martínez and Bosque 2013). In that case, on the basis of key principles of sustainable development, scenarios are analyzed in detail before back casting.

For detailed planning of any complex system, five hierarchical levels of the system must be distinguished, and the systems should be kept separate. These levels include: system level, success level, strategic level, action level, and level of tools. This model of five levels can be described as:

- Level 1. System (individuals, organizations, communities, nations in society, and biosphere).
- Level 2. Success in the system is based on basic principles (conditions of a system) to ensure social and ecological sustainability.
- Level 3. Success strategy: principles managing a process in order to implement sustainable development.
- Level 4. Specific actions that are a part of the strategic framework.
- Level 5. Tools (various tools and concepts to implement sustainable development).

The second level of success is the most important level in partner planning and cooperation with them. This level forms strategies, actions, and the choice of tools. The third level (strategy) is based on back casting in accordance with basic principles of sustainable development. For example, imagining that the conditions of success are assured and then asking what we should do now to optimize our ability to achieve the vision we are looking for and then asking what we should do next until the conditions of success are assured. The fourth level consists of specific actions needed to implement the strategy of success. Specific tools or combinations of them are chosen and analyzed in the fifth level. Tools that can be successfully applied in the implementation of sustainable development principles in an organization include: TNS methodology, GRI methodology, various management systems (ISO 14000, EMAS, SA 8000, etc.), assessment and monitoring systems based on sustainable development indicators, life-cycle assessment, Factor 4 and Factor 10, ecological production, concept of ecological footprint, natural capital methodology, etc.

Thus, it is necessary to create a specific policy orientated towards integrated environmental management instead of a segmental approach to it. This approach would include and evaluate:

1. the link among environmental factors at all levels (economic, social, and cultural) affecting the way mankind uses the environment and natural resources; the emphasis on renewable resources and a sustainable development;
2. the conservation and renewal of resources for a long period of time;
3. the preservation of the opportunity to use the environment and resources in an alternative way for future generations.

Ecological progress and economic welfare can go side by side, as ecologically adequate industry matches the requirements of the modern economy: economic usage of resources, development of waste-free production, refusal to use ecologically risky technologies, and the priority to the development of modern eco-safe technologies. As a matter of fact, a reorientation of the economy is inevitable. Businesspersons should understand as soon as possible that a healthy environment and a healthy economy go side by side. Manufacturers have to consider waste, its recycling, and utilization already during the production process.

## 2.3 THE ROLE OF THE COMPANY'S STAKEHOLDERS

Over the past decade, the stakeholder theory has become a major part of CSR development. CSR principles are important to company stakeholders and they want to exist in a sustainable country so it can be assumed that their role is an important reason why companies are implementing CSR principles. Stakeholders include all those that have an impact on the company and those affected by the company, so, according to these authors, the list of stakeholders is not finite; any person or surrounding environment can be a stakeholder that can also be divided even further (Morsing and Schultz 2006). Globalization has led to the creation of multi-national and transnational companies that continue to expand beyond national borders, resulting in a very broad geography of stakeholders; this means that stakeholders change not only in terms of quantity but also in qualitative terms, as each country's level of development, as well as the business environment, varies.

Scientists agree that two main groups of stakeholders exist: *internal* and *external.* This distinction depends on whether they have a direct or indirect effect. However, it is possible to distinguish three levels of CSR that affect the development of companies, i.e., *institutional, organizational,* and *individual* levels. These levels include certain dominant forces or, in other words, stakeholders. Institutional forces include state institutions, the media, and business unions, as well as trade unions (Pedersen 2006). These dominant forces can be treated as external stakeholders, but their impact is too minimal; thus, the internal forces of the company (internal stakeholders) can also be distinguished.

Nine stakeholder groups that are interested in the activities of the company can be distinguished (Table 2.4), together with their expectations.

A company's activities involve different relationships. The company, through its activities, impacts other groups (see Figure 2.4) and is itself affected by these groups and institutions.

Business has the greatest direct impact on its employees. Employees spend a large part of their lives in the organization. Therefore, what a company does or does not do is one of the most important factors in the life of an individual employee. On the other hand, business is highly dependent on its employees. However, the higher the number of employees in the company, the less direct influence each individual employee has. Nevertheless, every businessperson knows that employee readiness and the ability to work constructively can either promote better results in business or ruin it. The more sharing of ideas, friendliness, and communication there are, the more business relies on good relationships with their employees (Vanhamme and Grobben 2009).

Relationships with customers are also direct. These relationships are usually regulated by the market. It is very difficult for companies to ignore consumer needs and values. Only if the goods and services meet the needs of the users, and if they think that the company takes their needs into account, only then business is likely to

**TABLE 2.4**
**The Stakeholder Groups Interested in the Activities of the Company**

| Stakeholder groups | Expectations |
|---|---|
| 1. Consumers | Product quality and value, minimal consumption risk, good service and honesty |
| 2. Society | Employment, environmental protection, charitable activities, regional development |
| 3. Minorities | Equal rights and equal service opportunities, no discrimination |
| 4. Partners and suppliers | Long-term cooperation, regular and timely, fair billing |
| 5. Creditors | Favorable borrowing interests, credit repayment guarantees |
| 6. Shareholders | Large dividends, big capital gains, reliable investments |
| 7. Employees | Employment guarantees, fair payment for the work, work satisfaction, learning and training opportunities |
| 8. Managers | Adequate and fair wages, prestige, authority, career opportunities |
| 9. Authorities | The favorable tax system and good collection of taxes, the legitimacy of activities, compliance with normative legislation, reduction of corruption, development of democracy, development of social capital |

*Source:* created by authors.

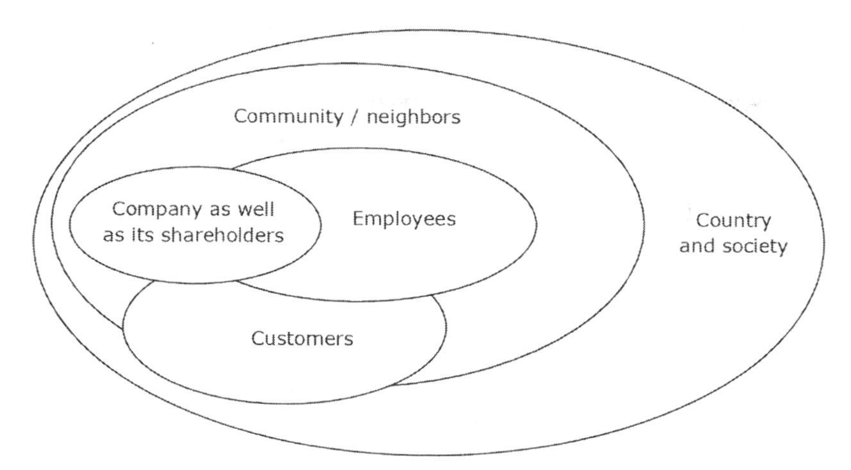


**FIGURE 2.4** Business and its stakeholder groups. (Source: created by authors.)

be successful. Since individual consumers are not inclined to analyze the characteristics of each product, taking into account its health and safety, and as the conditions under which the goods are designed and manufactured are usually not considered, the company relies solely on its reputation among consumers. As revealed by many public scandals related to conflicts about poor working conditions (in the company or between its suppliers), unsafe products, or contaminated food, reputation can be very easily damaged, and much effort is needed to recover and restore customers' trust in the company. Consumers are becoming increasingly aware of these factors; thus, in the competitive economy, they can choose what to buy based on their own interests and values. Increased publicity and awareness of the concept of social responsibility, as well as active publicity of socially responsible activities of the companies, increase the level of consumer awareness, while also raising expectations of the business community as a whole (Matuleviciene and Stravinskiene 2015). This is a business opportunity for all companies that can and want to separate themselves from their competitors by showing an example of socially responsible business development.

Relations with the community and society are not direct but equally important. Smooth relationships with the local community as a whole, and especially with local authorities (local government), depend on the goodwill and good reputation of a good business partner. Most of the business in the province depends on the local labor market opportunities when hiring new employees. Its reputation as an attractive employer makes it possible to attract talented people and to retain them in that region. This is also very important for the whole community. The importance of not polluting the nearest natural environment of the community is obvious.

Analyzing society as a whole, mutual influence is even less direct. While the overall national and societal model is more or less the same for all businesses, the positive or negative impact of individual business on society as a whole is limited and can only be felt at a certain level. This leads to "free user" behavior. In this regard, the business community as a whole and the specific business associations are strongly motivated to impact their members' behavior by avoiding the downward spiral movement and maintaining the level of the game field. Civil society (non-governmental organizations, consumer protection groups, the press, and other media) is responsible for the publication of social behavior, awards, and criticism on the anti-social behavior of individual businesses. By doing so, it creates fair and equal conditions for all businesses. Many researchers focus on analyzing the role and impact of primary stakeholders on companies. The very separation of the primary stakeholders already means that they are vital to the existence of the company and their values and expectations must be taken into account before anything else. This is why it is important to identify and assess what problems are significant for these stakeholders and how they can affect CSR. Although the views of different stakeholder groups on the desired business practices are quite different, a significant part of them have a common denominator and, thus, can be coordinated (Bailur 2006). A socially responsible company should identify the needs of each group and tailor its actions by choosing when and on which dimensions it should act. This statement would seem logical because different groups require different company responsibilities against each other.

Thus, the activities of each company have an impact on society, but first and foremost on stakeholders, some of which are closer, while others feel only an indirect impact. Each stakeholder group has its own interests (Clarkson 1995). A company that takes into account the interests of all stakeholders is considered responsible. The concept of CSR must be developed and promoted primarily in the interests of the business organization and other stakeholders (de Bussy and Suprawan 2012). Thus, business organizations, when designing their own social responsibility programs, must review existing social responsibility initiatives and improve as well as adapt them to the main concerns of their key stakeholder groups.

Three systematic levels of CSR can be distinguished: *macro, meso*, and *micro*. Macro factors promoting CSR are the influence of globalization, the promotion of national and supranational (EU) programs, as well as the influence of trade associations, labor organizations, and other external stakeholders. Meso factors include communities, society, consumers, and business partners. And finally, the micro level covers the organization's employees. It should be emphasized that, without some stakeholders, businesses would not even be able to exist (employees and consumers), which is why it is very important to meet their needs; otherwise, it can cause major problems and even threaten the viability of the organization. Thus, according to the theory of stakeholders, it is more important for organizations to establish long-term relationships with all stakeholders, and to adjust their interests, than to seek profit directly for the organization (Neville, Bell, and Menguc 2005).

Many authors have proposed a model that distinguishes three features: *power*, *legitimacy*, and *insistence* that stakeholders have:

- the principle of *power*: what power does the stakeholder group have when fulfilling their ambitions to influence the organization?
- the principle of *legitimacy*: are the actions legitimate, desired by society, and in accordance with its norms, values, and limitations?

- the principle of *urgency*: how urgently it should be reacted to the problem posed by the group?

According to these features, stakeholders are divided into eight groups as potential, self-dependent, demanding, dominant, dangerous, dependent, best, and independent.

It is necessary to highlight the importance of the stakeholders once more, since the potential risks are dependent on the potential impact of the stakeholders; thus, in order to reduce the risk of company management, social responsibility and commitment to all stakeholders are employed, as they help to identify the most important risks that arise. An important step for this is the identification of stakeholders by creating a power–interest matrix that helps in understanding the situation of the organization. Stakeholders with high power and strong interests are the most important to the company and the needs of this group must be fully satisfied. The needs of a stakeholder group with high power and weak interests must be satisfied sufficiently, but not that much of an effort is needed. Stakeholders with low power but strong interests need to be kept informed in order to balance their interests with the interests of the organization. And finally, those groups that have low power and weak interests should be only monitored by the company. If companies do not engage in dialogue with their stakeholders, favorable conditions for real threats and various risks are created (Donaldson and Preston 1995; Dickinson-Delaporte, Beverland, and Lindgreen 2010).

However, the greatest responsibility for promoting CSR falls on the shoulders of state institutions, as CSR is of particular importance to them since state institutions are a public administration entity that is in favor of the public interest. For institutional activities, everyone's, i.e., taxpayers', money is used, and therefore, they have to base their activities on effective mechanisms of the use of financial resources, efficient ways of managing the used natural, and material resources, as well as sustainable solutions that meet the needs of today, but do not diminish the ability of future generations to meet their own needs. Thus, as already mentioned, companies that use CSR principles contribute to sustainable development at the enterprise level, which requires state encouragement. In order to ensure the sustainable development of the country, targeted government policy is needed. In this case, the role of the state is to ensure the development of CSR and to create a favorable macro environment for the development of various CSR initiatives.

In pursuit of sustainable country development, states seek to promote CSR initiatives and use these means: subsidies, non-judicial, and non-fiscal tools such as awards and nominations, the spread of good practices, etc. The state can provide: subsidies for the creation of new work positions, tax exemptions for social enterprises, the possibilities to apply ecological, social, and other criteria for selection of suppliers in public procurement, the green procurement program, the possibility for the employee and the employer to be more flexible than foreseen in the laws when regulating mutual relations by collective agreements, etc. All these tools should provide the right macro environment for CSR initiatives.

Recently, the role of the public sector in promoting CSR has been prominent. First of all, public sector organizations support and initiate actions as well as exemptions that encourage businesses to act responsibly. However, the business incentive to be socially responsible should not be the sole objective of the government. In the context of promoting CSR, the government and its institutions can play certain roles at several levels.

The first level is the key role of the government: to shape the base environment. By applying different levels, the government has to define the minimum standards of business activity established in the legal framework, such as the regulation of working conditions by defining the maximum working time, occupational safety and health requirements, environmental legislation, etc. The government can take action by increasing the number of incentives and initiatives or by reducing the costs for companies, so they could make such changes. These would be specific government interventions in order to promote the dissemination of good CSR practices. Here the government or state institutions play two roles: facilitation and acceptance. Facilitation is treated as a variety of guidelines, funding mechanisms, and the creation of favorable conditions. This may include CSR promotion means, such as funding for research, publicity campaigns, dissemination of information, training, and awareness-raising projects. Public policy institutions can also promote the initiative market of CSR by creating tax incentives through investment and financing leverages. In addition, the government can promote the dialogue between all stakeholders at the national level to reconcile different interests. Political support and public sector support for CSR initiatives are very important. Approval can take various forms, from meetings to political documents or award schemes supporting innovative CSR initiatives, and directly recognizing the efforts of individual companies. The more developed stage is the fourth one, i.e., partnership-cooperation among stakeholders, and their mutual dialogue. In this way, public authorities have the opportunity to reconcile the roles of different stakeholders and to plan the policy direction, and solutions that encompass and integrate skills, which can complement their core competencies (Peloza, Loock, Cerruti, and Muyot 2012; Siriwardhane and Taylor 2014).

Successful completion of all stages also includes the category of demonstration (sample display) that includes actions such as ensuring fairness in the fight against corruption and increasing transparency of state institutions in relations with external stakeholders. In addition, public authorities must also apply CSR standards in their activities, for example, through the "demonstration" of the effect of public procurement, the organization of competitions, responsible use of resources, etc.

Based on this model of CSR encouragement by the state, it can be summarized that the role of the state is very important in promoting socially responsible business. In this case, a favorable legal framework for the operation of companies and other organizations can be created as well as tax incentives; extensive information on responsible activities can be provided, bilateral, tripartite partnership projects with businesses and NGO representatives can be performed, and guidelines can be prepared while awards can be organized.

Trade unions and civil society organizations can also stimulate business and apply CSR principles in their work by identifying problems, requiring the improvement of the situation and by searching for appropriate solutions together with companies. Consumers and investors, when choosing what to buy and where to invest, can increase market incentives for socially responsible companies. The media can inform about both the positive and the negative impact of companies. Thus, public authorities and other stakeholder groups should also be socially responsible in their multi-faceted relationships with organizations and businesses.

It is important to identify the main causes and obstacles to implementing CSR principles in the organizations that are discussed in the next section of the book.

## 2.4 REASONS FOR IMPLEMENTING CORPORATE SOCIAL RESPONSIBILITY PRINCIPLES

Companies are shifting from profit-seeking only for themselves to responsibility towards society and nature. They are becoming more sustainable and, in this context, the need for companies to take socially responsible action is emphasized. However, organizations do not want to be socially responsible if it does not benefit them in any way. L. Jakuleviciene, a UNDP representative in Lithuania, claims that social and environmental responsibility does not change the company's primary goal of profit. Therefore, socially responsible companies identify their stakeholders, the needs of these stakeholders, and seek to fulfil them, while at the same time trying to benefit from them.

While explaining the benefits of CSR, it can be stated that it gives companies a competitive advantage and leads them to economic success. If companies become socially responsible, they are able to gain more public approval, government support, capital, reliability, sustainability, and customer satisfaction.

It is worth noting that the European Commission strategy states that CSR principles benefit not only companies but society as well since when companies implement CSR principles in their activities, they contribute to the achievement of the EU objectives associated with sustainable development. CSR is associated with certain values, which can help build a more sustainable society and move towards a sustainable economic system. Social actions of companies improve the lives of local communities and reduce the need for state regulation, in this way creating better conditions for business. As mentioned before, most social problems cannot be solved without the help of business because it manages vast human and financial resources, some of which could be used to meet social needs.

In summary, two views can be expressed on CSR: for and against; however, nobody would want to be socially responsible if it did not benefit them. The majority of authors distinguish various benefits for companies; nevertheless, all of them lead the company towards economic prosperity and competitiveness, since CSR aims to accommodate and fulfil the needs of all stakeholders, thus avoiding conflicts and further problems. Properly fulfilling the needs of stakeholders makes the company employees and customers loyal to the company, while a positive image is formed; In addition, it helps gain public approval and financial benefits. However, it can be observed that not all CSR benefits are identified and described, and different benefits can be found and emphasized in each company according to the specifics of its activities. In addition, CSR brings benefits not only to companies but also to society and the country, helping to implement international commitments to reach sustainable development.

The organizational benefits and costs of implementing CSR principles can be grouped as follows:

- Reputation. Reputation consists of trust, quality, variety, consistency, and investment in human capital and the environment. Better corporate reputation helps attract and retain highly qualified employees.
- Competitive advantage. Companies and consumers share a strong connection; therefore, companies seek to promote the usage of their products in various ways to increase their output. CSR can help companies increase their competitive advantage.
- Employee loyalty. CSR is an important factor in enhancing employee loyalty and motivation.
- Ensuring stakeholder needs. Lately, companies have been facing increasing pressure from unions, employees, communities, and other stakeholders; therefore, CSR can help organizations meet their demands.
- Financial benefits. Investing in CSR can help improve the financial indicators of a company.
- Sustainable development. When companies invest in environmental protection, not only do they protect the natural resources on which both the business itself and population depend, but they can also improve the health of employees and all of society, increase their competitive advantage, reduce costs, create a market for organic goods and services, as well as promote innovations.

Various aspects motivate companies to implement CSR, such as the ability to better fulfil the needs of consumers, create a positive image, and build the reputation of a reliable company that would help increase loyalty, strengthen the connection with the local community and government, boost the sales of their goods or services, and raise employee motivation and work efficiency. This would improve human resources, reduce costs, and increase competitiveness (McWilliams and Donald 2001).

The majority of researchers have identified four main forces: the first one is the law; the second one is pressure from the stakeholders; the third one is economic opportunities; and the last one is ethical motives (the values of the company leader). The importance of the law is recognized by many other authors. Fines are increasing rapidly, so companies can lower their costs if they "beat the law to the punch." As mentioned earlier, the standards are becoming stricter, so it would be practical to "fend off" such regulations. Scientists have also emphasized the pressure from stakeholders. They claim that customers, local communities, and other stakeholders force companies to consider their environmental decisions. Being environmentally responsible, they avoid negative public attention and receive the support of the community. Another driving force identified in the model is the economic potential that improves the effectiveness of the production process, lowers the costs of proper waste management, etc. (Garriga and Melé 2004). And finally, the last driving force is that ethically motivated companies believe that the "right thing to do" is to be environmentally responsible. The values of the organization itself contribute to making it assess its role in society. In general, the model claims that the primary conditions to be environmentally responsible are created due to laws, better relationships with the stakeholders, the pursuit of economic prosperity and competitive advantage, and the ethical motives of the company. These driving forces behind corporate environmental responsibility can be compared with reasons which determine the choice of companies to implement CSR principles in their activities.

However, it can be noted that companies take on social responsibility in pursuit of certain goals. They take action and allocate funds for their own benefit since each company is primarily a profit-oriented organization. In conclusion, it can be said that organizations implement the principles of social responsibility to achieve different objectives because the benefits of CSR are numerous. Companies allocate funds with the purpose to satisfy the needs of stakeholders, but these costs eventually only benefit the companies in the long run.

There are various obstacles to CSR in different countries. According to the opinion of the Government of Lithuania, which is stated in the National Program for the Development of Corporate Social Responsibility for 2009–2013, approved in 2010, the development of corporate social responsibility in Lithuania is impeded by the lack of civic engagement and the weaknesses of unions and other non-governmental organizations. The development of corporate social responsibility depends on general public awareness and personal income. If a large part of the population is living in poverty and the middle class is shrinking, people have less money, are forced to save, and the criteria, such as the price of a product or services, start to dominate. Being able to pay more for the products of socially responsible companies and to support them in this way is only possible if people have greater financial possibilities. In the business environment, there is an opinion that corporate social responsibility is an expensive process without real benefits. In Lithuania, companies mostly compete on prices, sometimes on quality but rarely on reputation or social responsibility. The strategic objective of the National Program for the Development of Corporate Social Responsibility for 2009–2013 and its Action Plan 2009–2011 is to create the conditions for the development of corporate social responsibility and to encourage companies to implement the principles of social responsibility. The program has set the following objectives: to develop a legal and institutional environment which would support the development of corporate social responsibility, encourage a better understanding of corporate social responsibility, as well as social and environmental awareness, and would increase the competence of companies and stakeholders on the issues of corporate social responsibility. However, these measures and the inadequate financing for the CSR initiatives from the European Social Fund did little to contribute to the development of CSR, as most Lithuanian companies can barely survive due to the great burden of taxes and the growing shadow business, which further weakens the competitiveness of legitimate businesses. Therefore, instead of creating formal and imitation programs and wasting funds, the government should improve the conditions for business and reduce the tax burden on Lithuanian companies.

Tax evasion, contraband, and corruption are still tolerated in Lithuania, which in turn creates favorable conditions for shadow business. It is necessary to increase the number of taxpayers in Lithuania not only to fight shadow business, reduce cash transactions, strictly punish employers and illegally working employees, implement a general declaration of income, etc. but above all to remove the causes for shadow business. Reducing the tax burden and other bureaucratic obstacles for business can be a much more effective measure than all the other methods for fighting the consequences of shadow business.

The majority of EU countries encourage corporate social responsibility by implementing a public procurement and investment policy. Therefore, instead of allocating funds for raising public awareness and educating companies, the Lithuanian government should first work on reducing the tax burden on responsible business and improving the public procurement and investment policies by integrating the criteria of corporate social responsibility.

## 2.5 CORPORATE SOCIAL RESPONSIBILITY TOWARDS EMPLOYEES

The activities of a responsible business can be very diverse and different beneficiaries can engage in them. Corporate social responsibility can affect the following:

- promotion of staff learning;
- the organization of employee recreation, promotion of a healthy lifestyle;
- the increase of intergenerational solidarity;
- improvement of the psychological climate at work;
- development of equal opportunities;
- improvement of employee family well-being;
- environmental protection;
- promotion of transparent business.

In addition, socially responsible activities can be complemented by the following:

- assurance of occupational health and safety;
- implementation of environmental management measures;
- production and supply of ecological goods and services;
- participation in the activities of the community;
- promotion of the local economy;
- responsibility to consumers and customers.

A significant part of CSR activities is oriented towards employees, as a socially responsible company must take care of its staff. Therefore, employees are becoming an increasingly important factor in a successful organization. Researchers agree that the competence of employees is becoming a valuable source of competitive advantages. Consequently, a socially responsible company should first take care of its staff and increase investments into intellectual capital as well as ensure the health and safety of employees at work. Implementing these measures can secure a long-term increase of a company's competitive advantage, as it will help the company retain the best workers, strengthen their loyalty, and improve work satisfaction, which in turn will guarantee better performance.

Employees working in an organization perform certain roles which socially responsible companies strive to coordinate (Crane and Matten 2007; Dahlsrud 2008). Five employee roles can be distinguished according to their interests:

1. employees as people having a contract with the employer, contractual rights, and obligations;
2. employees as representatives of humanity with human rights and obligations;
3. employees as citizens of a nation with civil rights and obligations;
4. employees as people with the freedom of choice, ethical rights, and obligations;
5. employees as family members, which find family rights and obligations to be important.

These five roles define the protection of worker rights and welfare, as well as ensure that employees and employers fulfil their obligations (directly or indirectly). Thus, social responsibility to employees first comes from coordinating the interests of employers and employees.

Employees can be the main partners of a company in order to ensure that the publicly declared activities of the company are consistent with the actual activities. On the other hand, when effectively implemented, CSR impacts staff motivation, retention, and development (Elkington 2010). Employees come to companies hoping to use their opportunities and fulfil their needs. If a proper environment is provided to the employees, their commitment to the company increases. It is believed that there is a close connection between commitment to the company and socially responsible activities, focused on employees and fulfilling their needs. Financial gain is only one factor that motivates employees. Other motivating factors include a favorable working environment, positive relations with managers, a balance between personal life and work, the opportunity to learn and improve, a safe and healthy work environment, respect and human rights protection (Noel and Luckett 2014). Employees work better when they feel better, which can be achieved only by integrating certain elements in the work environment. This might include a positive micro-climate, the possibilities to have a flexible work schedule, to bring children to work, use comfortable kitchens, create good conditions for resting at work, and so on.

Socially responsible companies take these measures in relation to employees: creation of good working conditions, taking into consideration familial and professional commitments of workers; implementation of a flexible work schedule; employment of people with disabilities and other socially vulnerable people; promotion of lifelong learning; active communication between management and employees; implementation of an employee complaint system; a sexual harassment prevention policy; transparent wage payments (Husted and Allen 2006). As seen from these examples, the implementation of sustainable leadership principles plays an important part. By implementing the principles of sustainable leadership, socially responsible companies can not only increase the efficiency of organizational management and activities but also create the working conditions to make employees happy, which in turn would further improve the results, productivity, and work quality of the organization.

The function and duty of a socially responsible organization are to fulfil the various needs of its employees and to create good conditions. Therefore, while performing this function, the organization must ensure proper payment for work, create safe work conditions, provide the opportunity for employees to improve, and appreciate their efforts and achievements. The organization should also help employees solve problems which arise in their relations with other organizations (e. g., government institutions).

Consequently, organizations that want to attract and retain employees invest great funds into social programs for staff. This way, companies hope to secure the loyalty of this essential group of stakeholders. Employees are prouder of companies that find values and ethics important. Worker loyalty to such companies is stronger. Value-based companies that have a strong organizational culture find it easier to attract and retain employees, which has a positive impact on work results and financial indicators because clients are more satisfied with motivated employees. Thus, the implementation of CSR principles into the activities of a company can help improve relations with employees, ensure their loyalty, and increase their motivation to do the job well (Jamali, Lund-Thomsen, and Jeppesen 2015).

The implementation of the social-ethical CSR dimension in practice begins with the company's treatment of its employees. Socially responsible companies create the conditions for employee cooperation, teamwork, involvement in making decisions, learning, and self-realization. Socially responsible organizations also ensure equal opportunities, provide work variety, allow the workers to manage their personal and work life better, etc. Social initiatives aim to provide employees with more autonomy, improve their motivation, and create the feeling that work is meaningful while increasing their involvement in work and commitment to the organization. The latter aspect emphasizes the importance of the ethical treatment of employees in a company. It is beneficial for organizational managers to implement codes of conduct and ethics as well as to initiate training on ethical issues, as it helps create an effective company strategy and an atmosphere that promotes personal improvement. Employees become more motivated to reach the goals of the organization, they are proud of their workplace and feel a deep sense of commitment (Jamali and Neville 2011). Such workers create a positive image of the organization and improve its reputation, which further increases the competitive advantage of a socially responsible organization. However, the value system of the organization and the value systems of its members often differ. If the employees find the publicly declared company values incomprehensible or unacceptable, those values remain mere declarations. It is essential that the company and employee values and attitudes towards certain issues coincide or are at least similar. The coherence between company and employee values can be achieved through sustainable leadership, which would form a new organizational culture, based on the modern ethics theory and good experience of other organizations.

The following stages of forming social-ethical values of an organization can be distinguished:

- the initial stage, in which the employee group is entrusted to create the principles and concepts of socially responsible behavior;

- the specification stage, in which the definitions of socially responsible behavior, professional ethics codes, and special training plans are created;
- the integration stage, in which socially responsible behavior standards become skills, as they are integrated into the management of the company, its strategic goals, and core value system.

It is important to involve every employee in the integration stage so that the declared organization values are comprehensible and acceptable to everyone. The declared values also have to adhere to the values and attitudes prevailing in society. Only then does CSR not conflict with the interests of external stakeholders and instead becomes an effective instrument in the business competition.

Organizations themselves incorporate their social responsibility in ethics (conduct) codes, which the whole staff has to follow. They are one of the main internal instruments for implementing social responsibility in an organization. Although following the ethics code is voluntary, it plays an important role in developing the culture of the organization, shaping the values of its members, creating a positive attitude towards work and responsibility for its results (Kühn, Stiglbauer, and Fifka 2015).

The more the management of the company's staff is based on the principles and standards of ethics, the greater the likelihood that in such companies employee interests and expectations will be taken into consideration, opportunities to use their skills and knowledge will be presented to employees that will help improve their self-esteem, worker rights and dignity will be preserved, and the quality of life will be improved, which will contribute to the growth of human capital and human social development. The concept of human social development includes fulfilling both the basic (e.g., food, housing) and the complex (e.g., civic participation in social environmental processes) human needs. Therefore, organizations can contribute to human social development by primarily employing advanced staff management practices. The real-life examples of these practices are presented in Table 2.5.

Table 2.5 shows that companies can implement various practices in regard to employees, this way ensuring the protection of employee (social, moral) rights, forming relations based on dignity and respect in the organization, improving the quality of work and personal life, etc. Therefore, CSR initiatives should be important not only to company leaders but also to the entire staff, so that their implementation would be constructive. How the employees understand the social policy of the company impacts their willingness to participate, initiate, and contribute to social change initiatives. Workers want the company to act in a socially responsible manner not only because social responsibility lets them feel that the company treats and cares about everyone equally, but also because social responsibility initiatives require employees and managers to work together to reach higher goals. Employees also support social responsibility in order to improve the results of their activities and increase benefits. It is namely due to the decisions and activities of employees that social responsibility strategies are being created and implemented in organizations.

In conclusion, it can be claimed that a lot of CSR activities are directed towards employees, because they, as one of the most important groups of stakeholders,

## TABLE 2.5
## Advanced Human Resource Management Practices

| Practices | Examples of practices |
|---|---|
| Friendly/cooperative relations between the employee and employer | • responsible fulfilment of the duties and obligations of each party<br>• open communication between managers and employees (an "open door" policy)<br>• possibilities for an employee to express constructive criticism of a manager's decision (opinion)<br>• striving to reach a common goal<br>• mutual trust |
| Protection of employee rights | • the principle of equal opportunity<br>• physical and spiritual security<br>• freedom of choice<br>• fair employment, assessment, promotion, wages, termination, disciplining, and other staff management practices based on objective indicators<br>• promotion and organizing of learning<br>• adhering to health and work safety standards<br>• possibilities for employees to express their interests |
| Coordination of personal and work life | • child care centers in the organization<br>• covering the costs of child care and education (partially)<br>• purposeful use of employee time<br>• flexible work schedule, part-time work<br>• career continuation possibilities for employees on maternity (paternity) leave |
| Involvement in decision-making | • employees are consulted when making decisions concerning their work<br>• authorization to make independent decisions<br>• opportunities are presented to apply knowledge and skills at work<br>• causes of justifiable dissatisfaction with work conditions are corrected or eliminated |
| Practices that bring additional benefits to employees | • employee life insurance<br>• pension fund instalments<br>• interest-free credit for consumer goods<br>• free meals (especially in production companies)<br>• employees are granted access to the resources of the organization (phone, fax, car, etc.) for personal use<br>• promotion of a healthy lifestyle by covering gym and other similar expenses, organizing leisure and holidays |
| Positive psychosocial states of employees | • employees can feel free to express themselves and not to put up a front for each other<br>• dignity<br>• thinking that work is meaningful<br>• being proud of one's company or one's (team's) work<br>• feeling that the work climate is pleasant |

*Source:* created by authors.

determine all the main performance results of the company and are the basis for company competitiveness and successful growth. The implementation of sustainable leadership principles plays a crucial part here. By implementing the principles of sustainable leadership, socially responsible companies can not only increase the efficiency of organizational management and activities but also create the working conditions to make employees feel responsible for their work, enthusiastic, and happy, which in turn would further improve the performance results, productivity, and work quality of the organization. Socially responsible companies are committed to giving their employees more than is required by law, so as to secure the loyalty and dedication of this essential group of stakeholders. Social responsibility in regard to staff promotes employee responsibility and brings positive results to the whole company and every worker, as well as all of society because happy people who are satisfied with their work strengthen the solidarity of society and contribute to its safety and unity.

## 2.6 WORKPLACE QUALITY AND WORK SATISFACTION

Scientists have presented undeniable evidence that the organization's results are highly dependent on the happiness of the people who work in it. Experimental studies have found that happier people are more likely to be innovative, more successful in revealing their creative potential, and are characterized by higher labor productivity (Islam and Siengthai 2009). They tend to continually improve their skills and raise their qualifications; also, they do not spare energy and time at work and spread positive knowledge outside the organization, in this way enhancing a good image of the organization. Happier employees help to build and maintain an optimistic mood and efficient work climate in the organization. In comparison to unhappy employees, they are less frequently absent, have fewer harmful habits and better health, and perform the tasks assigned to them better (Sirgy, Efraty, Siegel, and Lee 2001).

Thus, the synergy between labor productivity and happiness is considered to be an important condition for employee motivation. It is recognized that investing in the quality of life at work can bring significant benefits to both employees and the whole organization. It is no coincidence that an increasing number of organizations are implementing measures of employee motivation that promote not only individual efforts but also teamwork relationships and a commitment to a mission that is consistent with public ideals, their role in the organization, and perception of responsibility, as well as the need to raise qualifications and to realize oneself as a personality at work.

The quality of human life at work is also closely related to the possibilities to combine different roles in work, family, and personal life (Haque and Taher 2008). Work satisfaction is enhanced by organizational support which helps employees to cope with the burden of these roles, to overcome the emotional tension caused by the abundance of tasks, responsibilities, and lack of time.

Improvement of the quality of work is relevant to both private and public sector organizations. The European Commission has submitted an assessment that, although the quantitative indicators of the EU's labor market are improving (employment is rising, structural unemployment is falling), concern about improving the quality of work exists, as there is some sort of "erosion" in terms of work quality: labor insecurity is increasing, working conditions are deteriorating (e.g., health problems linked to work-related stress has been increasing), while opportunities for combining work and private life are decreasing. The concept of workplace quality or quality of life at work has already been enriched by the contribution of psychologists at an early stage, first of all, the adaptation of the theory of needs by Maslow (Goudarzvand-Chegini and Mirdoozandeh 2012). Work satisfaction measurements or quality of life at work have been associated with meeting the needs of employees at different levels, starting from a material level, e.g., remuneration and sanitary working conditions, and ending with the highest reflections of personal life visions and self-realization. As quality of life at work became more popular, concepts were developed, new quality improvement factors were found, and recommendations for workplace improvement were provided. Initially, quality of life at work was related to work satisfaction; also, scientists suggested quite different sets of elements, based on theoretical and, very rarely, empirical research findings. Susniene and Jurkauska (2009) suggested identifying the essential components of quality of life at work according to external factors of work: remuneration, working hours, working environment, and an internal approach towards the work itself and its nature. Later, these aspects were complemented by the personal characteristics of employees, self-realization, participation in management, honesty and justice, social support, meaningful employment prospects, etc. (Cummins 2005). Such a pragmatic approach allowed the interests of employers and employee groups (e.g., trade unions) to be combined. Researchers of various disciplines who reflect on individual components of life quality have revealed the connections between various features of the workplace environment (e.g., internal motivation, involvement, the need of improvement) and satisfaction with work, which is an important indicator of the quality of life and happiness. Satisfaction with work is measured by methods which are used in health and psychological sciences by employing concepts of well-being, depression, and stress, as well as other concepts. For example, during the research on satisfaction, it was revealed that a significant connection between satisfaction and so-called "self-rated anxiety" exists (Juniper 2002). To measure the quality of life at work, the following indicators are proposed:

- adequate and fair remuneration;
- work conditions suitable for physical and mental health;
- use of people's skills and their development;
- personal growth, career, and security;
- space for the whole life (work, family, and leisure balance);
- constitutionalism (clearly defined employee rights and their protection);
- social integration at work;
- social responsibility (values of social organization, socially responsible behavior of employers and employees).

It is necessary to separate a specific work feeling—that is, people's emotions related to their work—and lifelike feelings of a more general nature, the so-called context-free well-being. This divide can be examined in three axes: (1) dissatisfaction–satisfaction; (2) anxiety–comfort; and (3) depression–enthusiasm. Such important work characteristics as physical security, socially valuable position, and opportunities to use skills affect employees' well-being at work. However, it is not only these factors that affect the well-being of employees. Some scientists believe that people's well-being at work also depends on personal attitudes, on socio-demographic factors (e.g., age and gender), and on the level of context-free well-being.

Researchers conducted a study on the relationship between happiness at work and labor productivity, looking for answers to the following questions:

- What makes people happy and what factors affect their happiness?
- How important is work to people and how does it affect their happiness?
- How do people assess their work and what kind of relationship does it have with their happiness?
- What is important to people when they are searching for a job and does it affect their happiness?
- Is there a connection among work, quality of work, and happiness?

Researchers also used the following indicators to measure people's work satisfaction:

- employee involvement;
- teamwork;
- commitment to the organization;
- work satisfaction;
- willingness to remain a member of the organization;
- positive working relationships.

Six indicator groups are used to measure employee satisfaction with work quality:

- remuneration;
- working hours;
- future perspectives (career and security);
- severity or complexity of work;
- work content;
- interpersonal relations.

As conducted research shows, even the highest amounts of money, if they are earned at detested work, bring neither success nor happiness. Moreover, such a person usually feels undervalued and it seems to him that his reward is poor in comparison to his "sacrifice." Thus, investment in the quality of life of employees at work brings long-term benefits to the organization. If you want to have happy employees in your organization, first you need to make sure that people are suitable for their work. Not only their experience but also their personal values, character traits, and interests should correspond to their work.

First of all, relationships with colleagues determine well-being at work (Kajzar and Kozubkova 2007). There are cases when the staff unites like a fist against a bad leader, but if the atmosphere is poisoned by co-workers' conflicts and intrigues, not even the perfect relationship with the direct manager will help.

A secure and trust-based environment is one of the most important sources of happiness at work. Few working people expect to maintain their workplaces for their whole life. Often organizations are destroyed, reduced, and reorganized; as even people with the highest positions may lose their jobs, the fate of ordinary workers is unclear. Such a persistent threat pervades people and they are afraid to take the initiative because it is related to possible errors enhancing the possibility of dismissal. The second thought that might come to mind and also paralyze the activity: why should I bend over backwards and work until I drop, if I could be fired any minute?

If an organization creates an environment in which everyone, irrespective of the economic situation, may feel supported as long as he/she tries, employees will work without fear and their contribution will always be higher. If people are treated fairly and openly, they can be relied upon in difficult times. However, talking about the sense of security, we should not forget that every stick has two ends. Absolute confidence in one's security encourages apathy, somnolence, and as such, the individual may cease to develop. This is as harmful as excessive stress. The dependence of employee productivity on stress is shown in Figure 2.5.

On the basis of various leadership theories, it can be said that the leader contributes to the creation of a secure, trust-based environment if he:

- encourages and supports initiatives, measured risks, and non-existence of fear of being wrong;
- is honest with his team members and allows them to freely express their opinion. He sometimes may ask for feedback face to face from his employees because they may be shy of their colleagues. If no feedback is

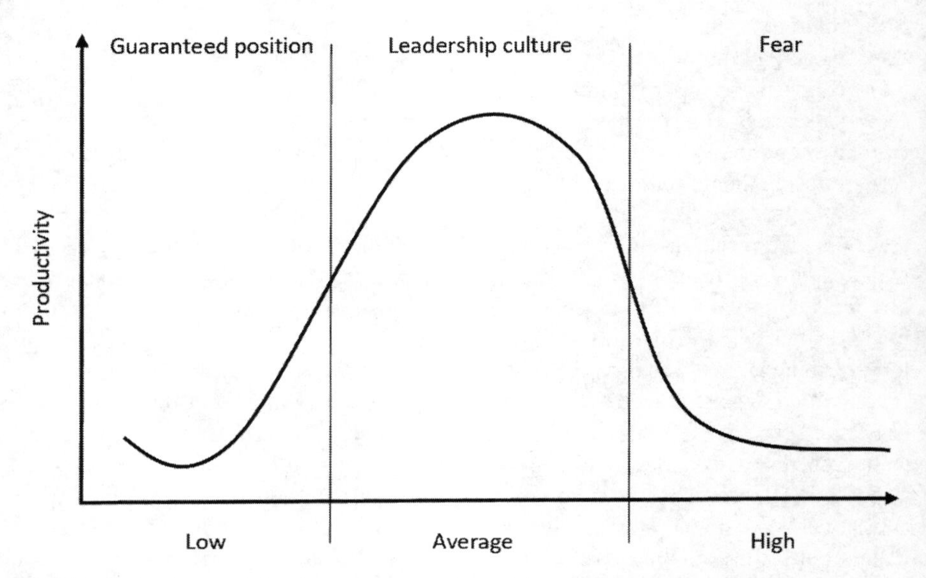


**FIGURE 2.5** Productivity dependence on stress. (Source: created by authors.)

received from employees, he is the ideal leader or there is no secure atmosphere in his collective;
- in response to criticism, remains positive and open; avoids excusing, justifying or rationalizing one's actions unless he is certain that the criticism is based not on his actions but on specific circumstances;
- gives each member of the collective the opportunity to be heard and allows everyone to express their opinion when it is needed;
- protects and supports his team members, if necessary—loyalty generates loyalty;
- does not threaten his own employees and uses the appropriate leadership style;
- declares an "open door" policy and ensures that this is indeed the case;
- devotes enough time for people to express their opinion in meetings with employees;
- avoids canceling planned meetings because people may feel less important in the leader's list of priorities;
- is certain that his deputies follow the guidelines listed above.

Trust in the organization is very important for the organization itself and it is up to the leader to have an atmosphere of trust. Scientists have examined this important aspect of leadership and offer a model that allows them to better understand what trust is and how it is created. There are three different categories of trust:

1. Strategic trust is the trust in the organization's vision, mission, and strategy, and its ability to exist successfully.
2. Organizational trust is the trust that the organization's policy will be honestly implemented, and the declared provisions will be respected.
3. Personal trust is the trust that subordinates show to their leader in the hope of his honesty and respect for their interests.

In a company, the most important aspect is to build personal and organizational trust. This reduces the spread of rumors and speculations that distract employees from work; it also motivates the employees.

Another important aspect is the involvement of employees in the planning and decision-making process. This creates mutual trust. Being able to control their work environment and managing their work, employees are naturally more self-confident (Susniene and Jurkauskas 2009). Increased self-confidence leads to higher self-esteem and such individual control creates a sense of ownership and initiative because it promotes employee creativity and innovation.

The listening skills are very important. Attentive listening to what the employee says and the response to his or her comments indicates that the employees are valued and respected. Specificity is very important when praising an employee. Employees are more autonomous and empowered if in the organization:

- it is recognized that the first step towards autonomy is not to constrain the contribution of employees to a common objective;

- it is demonstrated to employees that what he expects and especially appreciates goes beyond the established boundaries of the position. It is not allowed to give the impression that the employee is being asked to do more for the same salary;
- it is emphasized that the success of common activities mostly depends on each employee's personal initiative and effort;
- confidence in employees is demonstrated through words and actions; this helps to develop their independence;
- even if the employee who has been working by his/her initiative disappoints, the freedom to act is not immediately taken from him/her and there is no overreaction to his/her mistakes.

The benefits of educating and training people are evident and obvious. The better they are prepared, the more they can contribute to the overall result. Opportunity to learn is a self-value for many people who have ambitions to become leaders or just to climb the career ladder. If the manager is able to informally educate and train other people, even indirectly and next to his/her main functions, he/she gradually evolves from a manager to a leader.

Scientists distinguish several ways which help make the education process more successful:

- It is necessary to determine training needs, by aligning a person's role with the organization's expectations and not just by describing a person's position, but by expanding the list of features to be improved and incorporating the components of behavior, values, and attitudes.
- It is necessary to devote time to individual conversations about the employee's career expectations, and what can be done to help employees pursue their career.
- It is necessary to periodically review the progress made by employees in their development.
- Sharing your direct experience is very important; therefore, it is necessary to discuss issues with your employees and give them advice about how to deal with different activities or projects.
- It is necessary to involve people in different activities as much as possible even if it is not directly related to their positions.
- If the organization is unable to competently provide training services, it is necessary to contact professionals of this field.

Thus, a secure and trust-based environment and employee empowerment may be successfully ensured in the organization, particularly through the leadership principles. The top of the leadership pyramid, its fifth level, is the leader's personality. Very often the leader's personality is as important as his/her action. Some of the leader's characteristics are universally attractive and play a crucial role in attracting the followers of the leader. These characteristics may be developed and play a major role in becoming a leader. The combination of these important qualities allows the leader to find the emotional response of his followers, which encourages them to follow this person. The fifth and highest level of the pyramid, without a doubt, is much more complicated than all the other previously examined pyramid levels. It ensures the highest loyalty of followers and their aspiration to exceed their potential for a common objective. Sometimes this personality factor is called the leader charisma. Even observing children playing in the yard, we can say that some of them already have the characteristics of a leader. A leader's personality characteristics begin to develop from an early age. Some people lose them over time and others continue to do that for all their lives. House and Howell (1992) distinguish four key qualities of a leader's personality that are necessary to achieve the above-mentioned results: self-confidence; energy; empathy; beliefs.

*Self-confidence.* Of all the personality characteristics of the leader, this one is the most important. It counterbalances all the rest. Without this characteristic, all of the rest lose their value. No one will follow a person who does not trust himself. Following another person implies great respect and trust, and how can you trust a leader who does not trust himself? Man's self-confidence is what we immediately feel intuitively when interacting with him (Bernhardt and Osterman 2017).

Components of self-confidence in working activities could be considered:

- Perseverance in consolidating your opinion. It is the willingness and ability that the leader demonstrates fighting for his rights, beliefs, ideas, and values, as well as the ability to respect other people with this trait.
- Self-assured and calm posture, facial expression, and manners. All of this provides an unambiguous message to others about the leader's self-confidence and raises his authority. People draw conclusions about the leader's confidence from his body language, which includes posture and clothing, as well as manners of speaking and gesturing.
- Participation; the ability and willingness to engage in discussions of different levels. Being fearless in taking the risk of expressing your own opinion, which may differ from the usual approach, is an important personal attribute of the leader.

Human self-confidence also has an important internal aspect determined by how the person feels; it also has an external aspect, which defines how the person is understood by others. These two aspects are closely interrelated (Boxall, Purcell, and Wright 2007). Naturally, it is easy to show self-confidence to others when self-confidence really exists; however, the leader who shows it also increases the real self-confidence, although at first it may not be as strong as he would like.

Here are some examples of behavior that make you look more confident:

- Try to create an "aura" of confidence for yourself but at the same time observe that it will not turn into arrogance.
- Pay attention to your appearance, clothes, and accessories, create a "professional" image, and avoid actions that could diminish it.
- Observe yourself or try to get feedback from other people about your external effects.
- Do not hesitate to propose new ideas and try to share them with others.

- Never show your fear, although you may feel it; give as many bold ideas as possible.
- Do not give in to external pressure when defending your views and ideas.
- Always be tactful and do not exceed the limits of disrespect for others when defending your opinion.
- When you make a mistake, acknowledge that you are wrong.
- Do not try to undermine other interlocutors or colleagues; criticize others politely and reasonably.
- Express a critical opinion about other people only if they are there, by giving them the opportunity to explain or justify themselves.
- Always discuss reasonably, and provide clear, convincing arguments and references.

*Leader's energy.* When we talk about leadership, we perceive energy not only as physiological but primarily as a psychological thing. The lack of energy significantly reduces the impact of the leader on the people around him. Leader's energy is like a source from which everyone may draw energy.

Compound components of a leader's energy in work activities could be:

- positive stance or positive attitude and optimism in various situations coinciding with a sense of reality;
- enthusiasm, which is the spirit that leader may spread to other people and ignite them;
- internal engine or possibility to engage in physical and mental activities under any circumstances or external pressure.

The easiest way to become a more energetic leader is to start from the physical energy since it is much easier to train; in addition, a good physical condition and energy always have a significant impact on mental energy and working efficiency.

A few actions allowing others to perceive the leader as an energetic person are listed below:

- Always try to act with enthusiasm; such an attitude should be honest and not artificial.
- Try to encourage others as much as possible, especially the less energetic people.
- Actively seek opportunities and advantages in every situation and show them to others.
- Always remain optimistic and do not be overwhelmed by the negative attitudes of others.
- Never show your tiredness, find ways to control it; engage in a variety of useful practices to reduce tiredness and share them with others.

*Empathy.* Empathy is the ability to understand or feel what another person is experiencing. In the context of leadership, this means that a person can understand another person's attitude, not necessarily agreeing with that point of view. The basis of empathy is a sincere interest in other people and deep insight into their problems.

Compound components of leader's empathy in work activities are listed below:

- Active listening is an essential part of empathy allowing understanding of other people's values and attitudes.
- The ability to accept different opinions shows the leader's openness to diversity and demonstrates a broad approach to the problem.
- The interest in others shows a sincere desire of the leader to know what others can offer and what can be learned from it, as well as what kind of benefits may be provided to other people.

The only method to develop empathy is to try to look at different life situations in the eyes of another person. Therefore, empathy could be expressed as:

- an appropriate body language, showing the ability to listen and understand;
- attentiveness and listening while communicating with others;
- a sincere interest in the problems of others; questioning, refinement of details, and consideration of the problems of others;
- the lack of exclusive interest in one's own people and the lack of ignorance of others;
- openness to different opinions and points of view.

*Belief.* For many people, a belief in their own righteousness stems from the faithfulness to their values, attitudes, and formed opinions. The leader's belief in his righteousness, much like his energy, is the main thing that convinces his followers. This belief in own righteousness must be combined with flexibility and open thinking.

Components of a leader's belief in work activities could be:

- the reliability on or persistent faith in the correctness of the course chosen by the leader;
- the commitment or complete preparation to act, as well as unselfish devotion for activities;
- a delegation of commitments and strength to other team members.

Developing beliefs as well as other leadership qualities is quite difficult. For many people, a belief is associated with being honest to yourself. The internal belief is very important; however, how it manifests externally and how it is received and understood by the leader's followers is also significant to the leader.

A few actions allowing others to perceive a leader as a confident person are listed below:

- Never let the panic or other negative emotions break through; show your determination and firmness with your whole being, especially when you are experiencing difficulties.
- Give up your opinions and attitudes only if compliance with them is becoming pointless and you are certain that the members of your team have understood the reasons for this retreat.

- Never act in a way in which your actions could be interpreted as laziness; do not get away from any situation in which you are already involved, no matter how unfavorable it is for you.
- Constantly make sure that your words coincide with your actions and you are consistent and logical.
- Never openly show uncertainty about your set objectives or the results you want to achieve.
- Do not allow anyone to question the consistency and reasonableness of your attitude towards work.
- Say what you think, and always think about what you say.

The four above-mentioned characteristics of a leader (self-confidence, energy, empathy, and beliefs) give the individual the attractiveness and charisma in both personal and work lives but one of the most important features that distinguishes the leader from other leaders in the work environment is having a vision. This vision should, in particular, be linked to the creation of a sustainable organization. Only sustainable and socially responsible organizations can ensure the best quality of life at work for their employees since their objectives include broader goals than profit-making (Carré, Findlay, Tilly, and Warhurst 2012; Erhel and Guergoatlarivicre 2010). In this respect, as mentioned above, the social commitment of such organizations to their employees is particularly important.

The following chapter of the book presents the main insights of researchers and the authors of the book which are related to the concept of leadership and its interaction with management; in addition, the importance of sustainable leadership in implementing a vision of a sustainable organization is defined.

# 3 Management and Leadership

## 3.1 THE CORRELATION BETWEEN MANAGEMENT AND LEADERSHIP

While discussing the concept of leadership, it is important to note that it is often confused with the concept of management. Although the relation between the two concepts is relevant to this day and is still the subject of discussion, a general consensus has not been reached yet; there is no general agreement on the difference between leadership and management. Management, like leadership, has many definitions and interpretations. Various authors define management differently, but each definition contains the core idea that management is the guidance of a group (staff) towards the achievement of the organization's objectives (Cordona and Rey 2014; Boyatzis and McKee 2006). The majority of authors agree that leadership and management are not the same; however, they disagree on the correlation and main differences between the terms.

It can even be claimed that leadership is a type of management, a manager's ability to influence employees to work towards the objectives of the group with enthusiasm and confidence, while management is the ability to influence the group to achieve the set goals. However, making a clear distinction between leadership and management based on scientific sources is difficult, although it is indisputable that these terms have significant differences (Carnegie 2011).

The relation between leadership and management has also been widely discussed by Lithuanian authors. There is another problem in Lithuania: the translation of the English word "leadership" into the Lithuanian language. The English word "leadership" can be translated into Lithuanian in two ways: "lyderystė/lyderiavimas" (meaning leadership) and "vadovavimas" (meaning management). For this reason, the two closely related but different terms are often confused with each other in Lithuania. Therefore, it is necessary to define the differences between leadership and management, in order to determine the meaning of leadership.

Management is defined as the impact of the manager on the members of the organization, which is realized through the concentration and redirection of human resources to achieve the objectives of the organization. Leadership can be considered as the pinnacle of management, the ability to assemble the people willing and able to achieve the formulated objectives and the vision of the organization presented by the leader (Alberoni 2006). The power of management primarily comes from the formal position of the manager, while the power of leadership comes from the personality of the leader and the expectations of the members of the organization for this personality. In order to stress the difference between leadership and management, it can be stated that most organizations manage too much and do not lead enough. Usually, little attention is given to change, and not enough emphasis is put on the vision of the future, while managers primarily focus on the here-and-now of the organization.

According to scientists, leadership is akin to finding a path, and management is like going down an already established path. The following aphorisms exist: leadership is doing the right things, while management is doing things right. Managers do what is already well-known to them, do not seek innovation, and do not change anything. Leaders are the opposite: they are not afraid of innovation, promote change, and bravely accept challenges, essentially seeking to "find the right path." Therefore, leadership is primarily concerned with the changes in organizations, while management deals with difficult decisions and actions, facilitating the activities of the organization.

According to the famous US researcher P. G. Northouse, whose book on leadership is enormously popular in Lithuania (Northouse 2019), leadership and management have many differences and similarities: "when managers are involved in influencing a group to meet its goals, they are involved in leadership. When leaders are involved in planning, organizing, staffing and controlling, they are involved in management."

Thus, after discussing the concepts of leadership and management, it is apparent that drawing the line between the two is rather difficult. In many ways, leadership can hardly be separated from management, just as management cannot be separated from leadership.

In his famous book, published in 1990, Harvard University Professor J. P. Kotter (1990) stressed that it is necessary to differentiate the terms of leadership and management (Rajan and Ganesan 2017). He distinguished the following areas of differences between management and leadership (Table 3.1).

It is evident from the functions provided in Table 3.1 that leadership is a process associated with a change in the organization, while management is related to complex decision-making and control. Leaders primarily establish the courses of action, while managers plan and set budgets. Leaders usually inspire people, while managers control and solve problems.

However, in the literature, leadership and management are commonly presented as two separate concepts, differently affecting the development of the organization (Caldwell, Herold, and Fedor 2004). However, many functions and characteristics can be attributed both to the leader and to the manager of an organization. Although generally different people take these positions, the roles are closely related and intertwined. If management is considered the support of an organization, and leadership is its development, these two things cannot exist separately. It can be argued that there is a golden rule: not all managers are leaders, and not all leaders are managers. An effective manager must have the qualities of a leader, whereas an effective leader must also reveal his management abilities.

**TABLE 3.1**
**The Functions of Management and Leadership**

| Management ensures order and consistency | Leadership ensures changes and progress |
|---|---|
| *Planning and budgeting* | *Establishing courses of action* |
| • Create agendas | • Create a vision for the future |
| • Set schedules | • Explain the broader picture |
| • Allocate resources | • Determine a strategy |
| *The organization and staff selection* | *Uniting people* |
| • Provide the structure | • Present the objectives |
| • Distribute work | • Strive for involvement and commitment |
| • Implement rules and procedures | • Create teams and followers |
| *Control and problem-solving* | *Motivation and inspiration* |
| • Create ways of encouragement | • Inspire and excite |
| • Provide creative solutions to problems | • Empower subordinates and followers |
| • Take corrective actions | • Satisfy needs |

*Source:* created by authors.

Management primarily embodies the power of position, whereas leadership—the power of influence. Leadership occurs when someone attempts to influence the behavior of a group or an individual, while management, in turn, is a type of leadership, in which the pursuit of organizational objectives is a priority. Table 3.2 presents the main differences between leadership and management.

From the principles provided in Table 3.2, it can be observed that a manager is more oriented towards the current situation, while the attitude of a leader is more focused on the perspectives of the future; therefore, his actions and main criteria are directed not only towards solving the current problems. A leader is more oriented towards the common objectives of a team and the involvement of each individual in achieving those objectives; in addition, he is innovative and visionary. A manager, on the other

**TABLE 3.2**
**The Principles of Management and Leadership**

| Leader | Manager |
|---|---|
| Sets high expectations | Sets clear and measurable objectives |
| Determines the principles that help better understand priorities | Establishes the standards of activity |
| Looks for various alternatives to reach an objective | Organizes systems which allow for effective working |
| Prepares to solve future problems | Expertly solves problems |
| Observes from new and different perspectives | Applies past experiences |
| Strives to anticipate future needs | Takes action |

*Source:* created by authors.

hand, has clear objectives, as well as rules and standards on how to accomplish them, and is oriented towards processes and control. According to the majority of authors, leadership is a multi-faceted influence on relations, while management is a one-way relationship of authority (Kotterman 2006). Leaders are more concerned with the process of creating common objectives, while the main function of managers is to coordinate activities so that work can be done properly and on time. Accordingly, a manager administrates, and a leader innovates. A leader is the original, whereas a manager is a copy. A manager works based on control, and a leader—on trust. A manager does everything as required, whereas a leader does what is required. A manager mainly asks "how?" and "when?" A leader asks "what?" and "why?"

In most cases, leadership is expressed as a leading person's ability to unite people to achieve common objectives and is one of the essential parts of effective management. Therefore, leadership is primarily understood as a manager's competence to inspire and lead others in addition to successful management (Ardichvili and Manderscheid 2008).

To summarize the opinions of various authors, it can be claimed that the main function of a manager is to manage the organization here and now, to make sure that work is carried out on time and the objectives of the organization are achieved. Leaders are more oriented towards the future: they are concerned with the longevity of the organization and the success and results in the future. They strive to accomplish this by inspiring, motivating, and empowering their followers and teammates.

A person appointed to a management position is not always a true leader. Sometimes he does not have the qualities of a leader and in various situations does not act as a true leader would. According to P. G. Northouse, if a person who does not have the qualities of a leader is appointed to a management position, the organization is not successful and usually does not achieve good results. Considering the opinion of the mentioned author, appointed leaders can be divided into two categories: leaders and managers. The general opinion is that a person can be an impressive manager—a good planner and a great administrator—but he might lack the motivational skills necessary for a leader. Or he might be an impressive leader, capable of igniting enthusiasm and commitment in others, but might lack some skills of a manager.

In every organization, there are people who do not hold high positions but are universally acknowledged leaders. These established leaders do not always occupy managing positions in organizations. However, a person who is not in a position of management can become a leader through other members of the organization who support his behavior and agree with him. Personal qualities have the most influence over becoming such a leader: self-confidence, energy, enthusiasm, verbal activeness, awareness, willingness to know the opinions of others, as well as suggestion and assertion of new ideas.

However, currently, there is a tendency to regard the same people as managers and leaders. In the twenty-first century, it is not enough for companies to be managed only by managers; nowadays they need leaders, who think strategically and know what is and what will be important for the organization in the future (Eagly and Chin 2010). Nevertheless, in both the public and private sectors, not all managers are leaders. In addition, the differences between managers and leaders are quite striking (see Table 3.3).

**TABLE 3.3**
**The Main Differences Between Managers and Leaders**

| Managers | Leaders |
|---|---|
| Functionaries | Innovators |
| Acknowledge responsibility | Seek responsibility |
| Control employees | Trust employees |
| Competent | Creative |
| Specialists | Flexible |
| Set realistic goals | Set ambitious goals |
| Strive for a comfortable working environment | Seek challenges and an interesting working environment |
| Delegate with caution | Delegate and enable with enthusiasm |
| Perceive employees as hired people | Perceive employees as supporters |

*Source:* created by authors.

From the leader and manager descriptions presented in Table 3.3, it can be observed that the actions of a manager are predetermined by the rules set in the organization. Managers rarely experiment and do not seek challenges or innovation. Their aim is to organize employee work in a way that would help to effectively achieve the real objectives of the organization. Leaders, on the other hand, create their own way of working: they do not fear new challenges, but constantly seek them; they trust employees and are determined to work together with the team to adapt to constantly changing conditions as well as to overcome the emerging challenges. Leaders are particularly enthusiastic and creative, seeking and presenting broader objectives: they lead not only here and now, but also try to foresee and shape the future. Summarizing the differences between managers and leaders, we can come to the conclusion that those people who become successful leaders think systematically, trust their colleagues, are less oriented towards daily, traditional work, and more towards tendencies and forces that promote substantial change. They see and develop the vision for the future of the organization and inspire and motivate others to achieve that vision. This is of particular significance in the modern, constantly changing world full of challenges.

When comparing the functions of a traditional manager of the previous age and the modern manager-leader, it can be observed that the adaptation to changing conditions also modifies the importance of the qualities necessary for a manager (see Table 3.4).

When analyzing the main functions of a traditional manager and a modern manager-leader, a few significant differences can be observed. The modern manager-leader is the opposite of the traditional manager of the previous century. He is the erudite manager, open to innovation, concerned with both the organization and the staff; in addition, he is communicative, flexible, self-confident, and assertive. The modern manager-leader thinks globally, strives for diversified knowledge, is interested in all aspects of the organization, and is largely focused on the staff: appreciates the employees and understands that the best results can only be achieved by working in a united team.

## TABLE 3.4
## The Functions of a Traditional Manager and Modern Manager-Leader

| The functions of a manager | The functions of a modern manager-leader |
| --- | --- |
| Was the only one to make all the crucial decisions in the organization, to solve all the problems arising in the team he managed and played the role of an expert; | Shares responsibility with other group members; |
| | Helps employees solve problems; |
| Controlled the work process, took full responsibility for the work results of the entire group; | Encourages employee self-management and involvement; |
| Answered all possible questions, acted as an expert; | Raises important questions; ensures communication between employees and relevant specialists as well as feedback; |
| Created rules; | Forms a clear vision for the future of the group and organization, which unites the employees of the organization; |
| Valued staff consensus and the unity of opinion; | Appreciates people's skills and aspirations to express various opinions; |
| Aimed to avoid conflicts when possible; | Sees conflicts as opportunities to strive for synergy and improve the decision-making processes; |
| As a rule, reacted only to certain events, resisted changes; | Is active: initiates change and sees it as a key element for the survival of the organization; |
| Mainly focused on work objectives, production, and technical skills; | The main focus is not only on the work process but also on the people and results; |
| Used the linear, analytical style of thinking; | Thinks globally; |
| Aimed to gain specialized, functional experience and knowledge; | Erudite; strives to become a specialist in different areas and to get acquainted with various cultures; |
| Was only interested in the area of his expertise; | Interested in all the aspects of activities in the organization; strives to become a reliable partner to other groups and divisions of the organization; |
| Was in a heated competition against his colleagues; | Is a fierce competitor but at the same time strives for partnership with other competitors and sellers; |
| Worked only in the territory of his country; | Must be able to work on an international scale; |
| Perceived staff as a changing resource of the organization; | Perceives staff as the most valuable part of the organization because he knows how exceedingly difficult it is to replace a good employee; |
| Put the needs of the organization and not the employees first; | Strives to coordinate the needs of the organization and its staff; |
| Avoided risks; | Is prepared to take risks; |
| Functional style of thinking focused on a short period of time. | Systematic style of thinking, oriented towards long-term perspectives. |

*Source:* created by authors.

Most authors claim that inspiration, passion, enthusiasm, and emotions, which are common traits of the leader, are the opposite of the rationality, formality, and systematic nature of a manager (Adair 1998; Bolden 2007). The task of a regular manager is to carry out the mission of the organization by focusing on the main functions of management and to ensure effective work by emphasizing control. The task of a leader, on the other hand, is to create the right environment in which individuals or teams can take their own initiative to make the vision of the organization come to life. The leader is also the person who is followed voluntarily, while the manager is obeyed (Cooper 2005).

Trait theories claim that the best managers, who are capable of leading, have a combination of certain characteristics, such as high intelligence, wide knowledge, erudition, fairness, perception, initiative, a high level of self-confidence, etc.

An organization needs an effective leader and an effective manager to reach the desired result, but their contributions are not the same. In order to better understand people and their aspirations, gain their recognition and trust, ensure work stability as well as the clarity and importance of set objectives, a leader accepts new challenges and ideas and encourages changes that lead his team to a qualified completion of tasks. Therefore, management and leadership require people with different qualities.

Managers focus on the system and structure, while leaders focus on communication, motivation, and pursuit of objectives. Leaders often apply the "Seven S" (7-S) model, which includes structure (organization), strategy, systems, shared values, skills, style of management, and staff. This model is much more effective for leaders, compared to managers. Leadership is a process which seeks to create a vision for the organization and motivate the employees to aim for that vision.

A leader who strives to lead effectively must be able to perform two key functions:

- Solve various work-related problems.
- Seek to retain the group (solve conflicts, meet the needs of the workforce, encourage workers, etc.).

It can be noted that the function of solving work-related problems might be better attributed to management, not leadership, but authors often provide different views on the essential aspects of leadership and management (Gold, Thorpe, and Mumford 2010).

Successful managers-leaders have certain qualities:

1. Insight reflects that managers-leaders are managers who tend to win, for they always achieve what they want due to being able to clearly state their wishes and purposefully strive to realize them.
2. The ability to strategize describes the manager-leader quality of tending to achieve what he wants due to always having a well-thought-out strategy, which helps achieve set objectives.
3. Passion is inherent to a manager-leader, as his objectives also inspire others; leaders try to passionately work towards their objectives and at the same time inspire other members of the team

4. Fairness is characteristic to managers-leaders because they must demand integrity from themselves and others; therefore, this quality is an important and fundamental part of a leader.
5. Flexibility is an essential manager-leader trait, and managers who want to achieve the best possible results must be flexible: readily accept change, adapt to a constantly changing environment, as well as economic and social conditions.
6. Risk-taking reflects the manager-leader characteristic that helps them adapt to ever-changing conditions because good managers must always be prepared to take a calculated risk.
7. Team building is an imperative trait for a manager-leader, as building a good team that works effectively helps reach the objectives of the organization. A manager cannot do everything alone; he needs a united and supportive team.
8. Effectiveness reflects the manager-leader characteristic which expresses that leaders who are not afraid to take a risk are always well prepared to take action.
9. Setting priorities shows a fundamental manager-leader quality, that leaders select and present their priorities very responsibly.
10. The ability to control oneself is a crucial characteristic for a manager-leader, as it helps achieve better results in collective work.
11. Temperament is vital to a manager-leader because the temperament of the leader determines the liveliness, activeness, and energy of the manager.

According to some authors, the main difference between a regular manager and a manager-leader is the ability of the latter to foresee future factors: to have a clear vision and to choose the right strategies to achieve it. All of this determines the success of the organization. A leader does not overemphasize current events; he knows well where the organization is heading, and what it should do to be one of the leading companies in the future. A leader is open to innovation, while a regular manager is careful and afraid of it. Scientists present motivation as a crucial trait for a leader. A leader must motivate the members of the group, as reaching set objectives is only possible by working as a team.

Most authors are of the opinion that a non-leader manager is described in a highly negative way (Heskett 2007). A non-leader has certain negative qualities: emotional instability, impatience, incapability, glory seeking, slyness, arrogance, etc. We get the impression that an organization led by a non-leader is unsuccessful, as the manager seeks more benefit for himself, rather than the organization. It can be observed that, when a leader is discussed, the accentuated characteristics and attitudes are positive ones. A leader gets along with the team very well, respects his employees, does not control or persecute them without a reason, and is reliable, clever, understanding, and empathetic.

A manager-leader should primarily exhibit the following qualities:

- can be approached equally by every employee to solve not only organizational problems but also to provide personal assistance and support to team members;

- is actively involved in the management of the staff, implements progressive material forms of incentive, is personally acquainted with most employees, devotes a lot of time to converse with workers, is interested in their improvement, finding and selecting staff;
- does not tolerate a disconnected style of leadership, likes to interact with regular employees, to listen to them and to discuss their problems;
- is determined and tenacious, willing to assume responsibility, likes to take risks;
- is tolerant of open disagreement, grants authorization, bases relations with employees on trust;
- takes the responsibility for failures, does not waste time on looking for people to blame, as the main concern is to correct the mistake;
- is interested in the careers of the employees working in the organization, recommends co-workers for responsible positions first, and only in exceptional cases suggests outside candidates recommended by others;
- encourages independent employees, as the level of independence should fully correspond with the skills of a professional employee;
- does not unnecessarily interfere with the work of employees, controls only the final results and presents new tasks;
- has confidence in his own abilities, perceives failures as something temporary;
- constantly revises his work, seeks and implements effective innovations; therefore, under his leadership, the organization becomes mobile, is resilient to crisis situations, works effectively, and develops continuously.

The presented qualities show that a manager-leader is not only responsible and tolerant, but also active and diligent. In addition, the most important trait of a leader is intellect, which should be higher than average. A manager-leader must be able to think globally, quickly solve difficult problems, have initiative, and be independent and determined: able to voluntarily take action as well as encourage and motivate others to get involved. Broad erudition, knowledge in many areas, and the ability to navigate the international environment also play an important part in the activities of a manager-leader, strengthen his authority, and increase respect from others.

Therefore, a manager-leader first and foremost:

- has a vision for the future and creates it by himself;
- sets ambitious objectives and applies progressive and innovative methods to achieve them;
- is interested in the active work of the team;
- clarifies team objectives and delegates tasks accordingly;
- prepares, assembles, and empowers team members;
- knows the strengths and weaknesses of team members;
- clearly defines work areas and communication channels, ensures feedback;
- is extroverted, calm, reliable, determined, and diligent;
- is dominating but not oppressive, more enthusiastic and eager;
- has the most valuable quality of being able to attract, excite and, unite people;

- realizes his power non-aggressively, by respecting the needs and feelings of others, and listening to the opinions of the members in the team.

By reviewing the personal qualities of a leader presented by various authors, it can be determined that the most essential qualities of a manager-leader are reiterated: intellect, the pursuit of responsibility, self-confidence, pursuit and implementation of innovations, provision of freedom for employees when coordinating their work, etc. These traits distinguish managers-leaders from non-leaders or regular managers (Ibarra 2012; Lawrence 2010). However, a question arises: do they determine the success of modern managers-leaders? Do all successful managers-leaders express these qualities? These questions will be answered by practical examples of sustainable leadership, which will be discussed in the last chapter of the book. The main characteristics of managers-leaders and the results that they have achieved will be presented there.

## 3.2 STYLES OF MANAGEMENT

Leadership and the style of management are closely related, i.e., the combination of personal qualities, values, and priorities determines whether the appropriate style of management is applied or rejected. In most cases, leadership is expressed through the manager's ability to unite people for the achievement of common objectives and is considered one of the crucial parts of effective management (Rothstein and Burke 2010). Leadership is seen as the manager's competence to inspire and lead others in addition to successfully managing them. Leadership competence is primarily influenced by the objectives and specifics of activities within the organization. The various management styles associated with the specifics of the organization are presented below. Not all management styles are suitable for managers-leaders.

*Autocratic management.* This type of management is characterized by the manager having absolute authority and making all the decisions on his own (Whitmore 2002). This management style is especially useful with employees who need close monitoring while performing certain tasks, and who cannot work without constant control and supervision. Creative employees who prefer teamwork resent such management, as they cannot change processes on their own or make the necessary decisions. This causes dissatisfaction with work. This style of management is inappropriate for a manager-leader, since, as mentioned before, a leader is usually followed and not obeyed.

*Non-intrusive management.* A manager who applies this style of management does not provide constant supervision, because highly competent, conscientious, and responsible employees do not require continuous supervision and control to achieve the necessary results. This management style is related to the definition of leadership, as managers-leaders barely manage and do not supervise team members. This can cause a lack of control and higher costs for the organization due to the provision of unsatisfactory services and missing of deadlines, as not all employees are responsible and can work without any control (Castro 2013). In order for employees to perform their duties perfectly and without control from the manager, a certain organization's culture level is needed, which is not so easy to achieve, especially in post-communist countries. Therefore, an organization's culture is one of the most important factors allowing successful implementation of management based on leadership.

*Democratic management.* A democratic manager listens to the ideas of his team members but makes the final decision by himself. However, team members gain a feeling of satisfaction and self-worth due to their contribution to the final decision. Such a style of management is suitable for a manager-leader. This management style, based on the involvement of team members, helps the team adapt to changes better than any other type of management, as people know that they have been consulted and that they have participated in the decision-making. Therefore, resistance and intolerance in the group are avoided or reduced to a minimum. This management style is disadvantageous when making decisions quickly or at any given moment, as there might not always be enough time to discuss everything with the group and make a collective decision (Jung 2001).

*Bureaucratic management.* Such management is very structured, and the manager strictly adheres to established procedures. This type of management has no room for new methods of problem-solving. These managers ensure that all the specified steps and procedures are performed before passing the matter on to the sphere of higher competence. Universities, hospitals, banks, and governmental institutions usually require this type of management to ensure quality and safety, and to reduce corruption.

*Charismatic management.* A charismatic manager manages by energizing and inspiring his team members. Such a manager can be considered a leader (Yammarino 1999). He should commit to the organization for a long time. If the success of a project or work is attributed to the leader but not to his team, a charismatic leader might pose a significant threat to the company by deciding to leave and to take on new challenges. In this case, the employees of the company would need a lot of time and hard work to regain their self-confidence and get used to a different type of leadership.

*People-oriented management.* This type of manager supports, instructs, and trains his staff, increasing employee satisfaction and interest in work, in order to achieve efficiency and productivity. This management style is also fitting for leaders, because supporting and training staff, and providing them with optimal working conditions, can ensure staff loyalty and commitment to the organization, willingness to work effectively as well as to strive towards the objectives set by the leader, and to personalize the said objectives (Ryan and Tipu 2013).

*Task-oriented management.* A task-oriented manager focuses mainly on work and specific tasks assigned to each employee. This style of management suffers from the same motivation problems as the autocratic management style, which does not consider the needs of the team in any way. Achieving results directly depends on close employee supervision and control. Therefore, this form of management is not suitable for leaders.

*Serving management.* A serving manager helps to pursue an objective by providing his employees with what they need to increase their productivity. Such a manager becomes an instrument used by employees when pursuing an objective but is not a managing voice. This style, in a sense, is similar to democratic management;

however, the results are achieved over a longer period of time, although staff commitment is even higher. Serving management is appropriate for a manager-leader, as it lets him share the responsibilities with the members of the group or organization for the sake of a common objective.

*Effective management.* An effective manager has the power to perform certain tasks and reward or punish the team for its work. This power provides the manager with the opportunities to evaluate, correct, and teach subordinates, when their productivity is below the desired level, and to reward for effectiveness when desired or better results are achieved. Effective management can be efficiently employed by managers-leaders.

*Change management.* A change manager motivates his team to be effective and efficient. Communication is the basis for the pursuit of objectives. This type of manager is clearly visible, and finds the overall picture the most significant, while the people that surround him take care of the details. Such a person always seeks new ideas to help achieve the vision of the company. This management style can be effectively implemented by managers-leaders, as it allows them to inspire and motivate team members to pursue a vision, which encompasses important changes.

*Environmental management.* An environmental manager is the one who knows how to create the right organizational environment and, in turn, impact the emotional and psychological perception of a person's place within it. In order for this style to be effective, it is important to understand and adapt the group psychology and dynamics to the organizational environment. Such a style of management uses organizational culture to inspire individuals and to develop new leaders. This management style is appropriate for a manager-leader, as it lets him achieve good results, involve team members, and inspire them to strive for a common objective.

However, the main traits which separate managers-leaders from regular managers could be the following: leaders consider the long-term perspective; by thinking about the whole situation, they understand relations better; they influence the components of the organization outside of their own area of command; they mostly emphasize such intangible things as a vision, values, and motivation; they intuitively understand irrational instinctive elements of interaction between the leader and other units; they are able to skillfully coordinate conflicting requirements from numerous organizational components; they continuously contemplate innovation (Elloy 2005).

People can be trained to become leaders (Samad 2012). Self-creation is great human power. A leader is a person who, based on his own experience and world view, finds and provides the functionally correct decision for action with regard to social gestalt (which he is a part of himself), and gains a moral reward together with a spiritual expression for it. Leaders express themselves as personalities and achieve much more than a regular manager. A broad spectrum of knowledge is not enough for a leader. A leader must comprehend and integrate this knowledge to some level of consciousness through his own understanding and vision. A conscious leader has to comprehend complex knowledge through all forms of wisdom: intellectual, emotional, and spiritual. But first, it is necessary to know that such forms of wisdom (intelligences) exist, as well as to understand their origin and the possibility to develop and use them.

**TABLE 3.5**
**The Main Elements of Human Intelligence**

| Capital | Intellect | Function |
|---|---|---|
| Physical | PQ physical intelligence | What is my physical condition? |
| Intellectual | IQ rational intelligence | What am I thinking? |
| Social | EQ emotional intelligence | What am I feeling? |
| Spiritual | SQ spiritual intelligence | Who am I? |

*Source:* created by authors.

The main elements of human nature are the body, mind, heart, and spirit. Every person has four types of intelligence or ability corresponding to these elements: the physical (body), rational, emotional, and spiritual (Table 3.5).

To summarize, it can be claimed that an individual can be a good manager, i.e., a good organizer and planner, but he might lack some qualities inherent to a leader. On the other hand, there are some successful leaders who are able to motivate people but lack skills characteristic to a manager. Either way, management must be effective, and effectiveness requires various skills and different styles of management.

Most modern organizational managers are not perfectly prepared to be leaders; thus, they apply the wrong styles of management, which oppose the concept of leadership (Esigbone 2000). When analyzing the activities of such managers from a professional aspect, they seem to know their activities well enough but are often unable to solve important social and psychological problems in certain situations.

It is essential to admit that the future of leadership should be associated not with solving problems of efficiency or control, but with increasing the accomplishments of people and achieving improvement. Business companies seeking to claim and retain leading positions in the market should primarily strive for unity, seek to create a proper organizational culture, implement corporate social responsibility (CSR) principles, and trust managers-leaders, because organizations will be unable to unconditionally trust only obedient and short-sighted managers-doers in the future. It is sensible and beneficial to focus on managers that have the potential for leadership, and to train them, since they are the ones bringing success both to the business and to the employees of the organization; in addition, they make the organizations sustainable, which, in turn, directly contributes to solving crucial social problems and implementing sustainable development principles in Lithuania.

## 3.3 THE CONCEPT OF LEADERSHIP AND THE BASIC TRAITS OF LEADERSHIP EXPRESSION

Leadership emerged in the English language a few hundred years ago and subsequently passed into many other languages. The conception of leadership process is based on the assumption that, at different times, one or a few people can be identified as leaders according to their obvious features which distinguish them from other people, subordinates. As already mentioned, disagreements arise not only because of the leadership concept but also because leadership is often confused with management. First of all, disagreements over the definition of leadership arise due to the fact that it involves the interaction among the leader, his followers, and the situation (Essien, Adekunle, and Oke Bello 2013).

Some researchers portray leadership by using personal and physical characteristics, while others describe leadership as behavioral traits; still, other scientists argue that leadership is a temporary role, which may be occupied by any person. However, among all these descriptions there is one common trend—social influence.

A short definition of leadership could sound as follows: *leadership* is a process of social influence in which the leader seeks the voluntary participation of subordinates, in order to achieve common organizational objectives. A more formal definition is provided by the GLOBE Research Group (House, Dorfman, Javidan, Hanges, and Sully de Luque 2014): leadership is the individual's ability to influence, motivate, and enable others to work for the organization. As can be seen from this definition, leadership means more than power and authority and is divided into different levels. For example, at the personal level, the concept of leadership includes advice, training, inspiration, and motivation. Leaders bring together teams, create relationships among group members, and deal with problems at the group level (Mester, Visser, and Roodt 2003). Finally, leaders create culture, at the same time making changes at the organizational level, which guarantees the best results for both the company's management and the development of the specific organization.

Thus, leadership is primarily viewed as a process where one person influences others and, during that process, a certain group of people achieves a common objective while acting together. Other scientists suggest defining this concept as the process of leader's impact and its guidance of group members' activities, required to accomplish the task. Hence, four important conclusions, which are discussed below, are drawn from this definition.

Leadership involves other employees or followers. With their determination and willingness to obey the leader's orders, the group members help to anchor and define the status of the leader as well as to set out conditions for leadership.

Leadership means unequal power distribution between the leader and group members. Group members have some power: they can shape, and indeed structure, the activities of the group in various ways, but the leader usually has more power. The manager has five powers: reward, coercive, legitimate, referent, and expert powers. The more of these power sources the manager has at his disposal, the greater his potential of effective leadership is. However, in the activities of organizations, it is becoming increasingly common for the managers of the same level and with the same formal power to differ in their abilities to employ reward, coercive, referent, or expert powers.

One of the leadership aspects is the ability to use different forms of power, making an impact on followers' behavior in various ways. The successful use of a leader's powers focuses and motivates group members to get involved, commit to, and achieve the common objective.

The fourth leadership aspect combines the first three leadership aspects and recognizes that leadership is primarily focused on values. Righteous or moral leadership requires values to be taken into account and sufficient knowledge to be given to the followers about the alternatives so that they themselves can make conscious choices when the time comes to decide whether to follow the leader or not (Nwadukwe and Court 2012).

In the majority of leadership concepts, four main aspects are mentioned: influence, power of personality, group, and objective. Thus, leadership is related to the influence: it is very important how the leader affects the followers since influence is a mandatory requirement for leadership. Therefore, leadership can be defined as the process by which a person influences a group of people to achieve a common objective. Leadership always occurs in groups. Groups are the environment for leadership. The leader helps to achieve a common objective by influencing all members of the group. This influence is exercised through both the power of personality and the application of innovative leadership approaches based on inclusion as well as various principles and practices of sustainable leadership.

There is a variety of leadership definitions, but they all consist of the same components, such as:

- Leadership is a process.
- It is based on values and new ideas.
- It covers the influence of a leader.
- It takes place in a group context.
- It includes achievement of an objective.

In all leadership theories and concepts, these components play a certain role in the definition or the applied theory. Leadership is also associated with the achievement of objectives; it is a process where the leader must connect a group of people for the performance of a particular task and the achievement of a specific objective. Thus, leadership emerges and creates a means of influence in an environment where people seek purpose. The best leaders create the best results, in particular, through their influence on followers.

Figure 3.1 presents the essence of leadership.

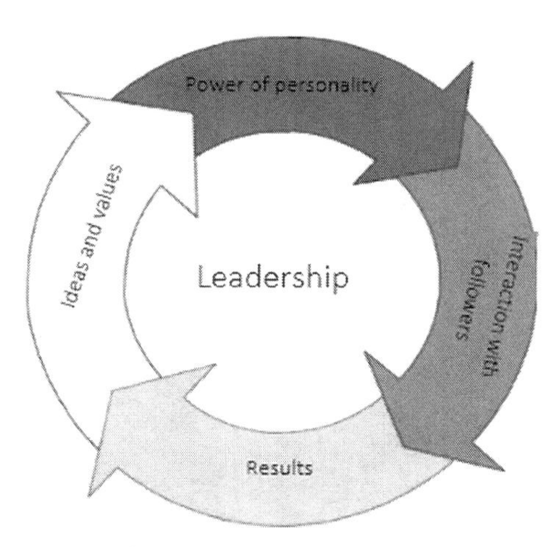


**FIGURE 3.1** The key components of leadership. (Source: created by authors.)

As is apparent from leadership components provided in Figure 3.1, leadership is the ability to bring people together and give meaning to their activities, through the power of personality as well as through ideas and values, in order to successfully implement objectives and achieve the desired results.

Therefore, the key leadership components are the following:

1. Vision, values, and ideas (value orientation, moral characteristics, strategic management; openness to changes, innovations, etc.).
2. The power of personality (self-confidence, ability to inspire, strong moral values, image, communicability, intelligence, wide erudition, good sense of style, etc.).
3. Interaction with the followers (cooperation, orientation to people, teamwork; communication, feedback; conflict management, sustainable management, etc.).
4. The results (goal orientation, directivity, the pursuit of an objective, etc.).

In the next section, the main leadership theories and leadership types, emphasized in those theories, are discussed. In one way or another, all discussed leadership components are highlighted in different leadership theories and included in leadership definitions (Westwood and Chan 2002).

## 3.4 THEORIES AND TYPES OF LEADERSHIP

Leadership theories are directly related to leadership concepts applied by authors (Duknytė 2015). The majority of researchers have chosen to examine and analyze one aspect of leadership: personality traits, behavior, relationships with subordinates, etc. The course of leadership research can be described in three stages:

1. Trait theories. They are based on identification and description of characteristics usually found in leaders. These theories assume that the organization will perform better under the management of a person who has specific characteristics of a leader and that leaders stand out among other people because of their special characteristics.
2. Behavioral theories. They identify traits of behavior that are essential for a leader. The most important aspect in these theories is not who the manager is but how he behaves and what management styles he applies.
3. Theories of the interaction and influence between the manager and followers. Based on these theories, a leader is a person who has a vision and leads subordinates by showing them the right way.

Trait theories state that some individuals have special innate characteristics or skills that make them leaders. Leaders are characterized by unique physical features, personality traits, and skills. Based on the trait theory, the concept of leadership is formulated as a set of features specific to different people. According to this theory, leadership is only pertained to the chosen ones that have special innate talents or have developed characteristics specific to leaders. Trait theory describes charismatic, authentic leadership and applies psychodynamic leadership models (Goleman, Boyatzis, and McKee 2007). Some sources also include theories of processes. According to them, it is believed that anyone can become a leader if leadership is identified as a process. Leadership as a process can be monitored in the behavior of leaders and this behavior can be learnt.

Manager behavioral theories focus on the behavior instead of who the manager is. Subsequently, theories of contingency emerged, which analyze how specific situations or conditions change the behavior of the manager and how it impacts his management style. In these theories, concepts of leadership based on visions, and moral and spiritual values were used.

In the middle of the twentieth century, the versatility of characteristics attributed to the leader was questioned; however, personal factors related to the personality of a leader remained important in further research. The Modern Skills Approach states that effective leadership requires not only features but also specific knowledge and skills. Here, leadership is analyzed as a set of developed specific skills. Skills include knowledge and experience that can be learnt or developed, unlike personal characteristics. According to the Skills of an Effective Administrator theory, effective leadership depends on three main groups of skills: technical skills (competence in a specific area, analytical skills, and ability to work with specific tools or technologies), human skills, and conceptual skills. Leadership style theories are also essential. Style theories generally focus on the behavior and activities of leaders. Researchers of leadership styles have established that leadership consists of two types of behavior: task-oriented and relationship-oriented. Task-oriented leadership allows the objective to be achieved more easily, whereas relationship-oriented leadership allows subordinates to feel more comfortable with themselves and when interacting with each other in current conditions.

In the last stage of leadership theories, the interaction and influence between the manager and his followers are explored. Based on these theories, a leader is a person who takes the lead and shows the right way to his followers. At the beginning of the 1960s, Blake and Mouton (1966) examined the behavior of organizations' managers focused on the task and the relationships with employees (Haris 2010).

Summarizing, leadership theories can be divided into two separate groups:

- Instrumental theories (theories of traits, management skills, the behavior of leadership styles, and situational leadership) that emphasize the behavior of the leader, his knowledge, experience, and characteristics that help the leader to manage more effectively.
- Inspirational theories (social interaction and leader-member exchange theories, social influence, transactional and transformational leadership theories, as well as resonant or emotional leadership theory, and the New Paradigm or the Future of Leadership theory) that emphasize the vision, values, and the ability to motivate themselves as well as others when aiming for changes.

Based on leadership theories, the most widely described leadership types, grouped according to common elements of leadership presented in Figure 3.1, are discussed below (Table 3.6).

**TABLE 3.6**
**Types of Leadership**

| | |
|---|---|
| POWER OF PERSONALITY<br>Character orientation | Charismatic leadership, authentic leadership, psychodynamic approach of leadership |
| INTERACTION WITH FOLLOWERS<br>Orientation to people | Transactional leadership, transformational leadership, transcendental leadership, shared leadership, distributed leadership, servant leadership, emotional or resonant leadership |
| VALUES, MISSION, AND VISION<br>Value orientation | Visionary leadership, value-based leadership, mission-driven leadership, spiritual leadership, moral leadership, sustainable leadership, New Millennium leadership or New Paradigm leadership |
| RESULTS<br>Orientation to an objective | Results-based leadership |

*Source:* created by authors.

*Charismatic leadership* is the oldest, most easily perceived and popular type of leadership that is very often mentioned in scientific sources. Charisma means characteristics of an individual, the ability to influence people, inspire global enthusiasm, and receive approval. A charismatic leader is a person who has unique abilities to inspire others, influence them, and achieve unbelievable results. Such a leader is respected, admired, and trusted, and is capable of creating a positive environment and good microclimate in an organization. In addition, such leaders distinguish themselves with strong values, tendency to dominate, non-traditional behavior, desire to have an impact, ability to inspire others, great self-confidence, high self-esteem, and great authority. A charismatic leader has great confidence in himself, shows an example, clearly identifies objectives, has high ambitions, inspires his followers, is competent, communicative, and attractive, as well as having a great image and reputation. It is the person you want to resemble or at least try to emulate. That sort of a leader has a tremendous impact on his followers, who genuinely trust him and do not critically assess him. His followers unconditionally agree with him, obey him, get attached to him, expect to receive help from him, and at the same time become more self-confident, have higher ambitions, as well as develop and grow. By helping their followers to gain more self-confidence, charismatic leaders expect better performances, as work will become followers' tool of self-expression, their source of happiness, and satisfaction with life. This approach helps employees to relate more to the organization, to strengthen their identity, sense of commitment, and loyalty, and improve their well-being, as well as create a better general microclimate within the organization. In addition, charismatic leaders are persistent and not afraid to take risks; they pursue objectives in unconventional ways, set an example but do not seek consensus, show self-confidence, pay attention, and show compassion for their followers. Their power as specialists and professionals is acknowledged even though their position may not grant it. This brings success to both leaders and their followers. It is significant to notice the importance of feelings in communication between leaders and their followers. Usually, a few examples of a charismatic leader are provided: this is a person with high aspirations, moral implications, well-grounded high self-esteem, and dependence on his country. Charismatic leadership is very effective because of the links between followers and their beliefs with the leader and the organization. This leadership is useful when there are huge restrictions, people are tired of pressure, and are waiting for a leader who will save them by solving or at least taking on the problems. Such leadership is best suited for short-term activities and small projects; however, it is less suitable for continuous activities that require a lot of work and patience. In addition, charismatic leadership is criticized for possible destructive objectives as, after applying it, a possible danger for an organization occurs. Due to possible mistakes or resignation of a leader, an organization might experience failures, loss of enthusiasm and hope. Since the success of an organization depends on the leader, followers usually obey him unconditionally, which means that they do not play an important role themselves and their development is limited. There are examples of charismatic leaders pointing people in the wrong direction. Lately, charismatic leadership is approached cautiously because of its focus on a single person, which results in a loss of attention to the community, universality of activities, and meaningfulness of activity of every employee.

Examples of charismatic leaders could be Martin Luther King, Angela Davis, Presidents Nelson Mandela and John F. Kennedy, Mahatma Gandhi, Pope John Paul II, etc. Nevertheless, there are also negative examples of charismatic leaders, such as Adolf Hitler, Benito Mussolini, Napoléon Bonaparte, etc.

The concept of *authentic leadership* was formed in the twenty-first century. Authenticity is guided by values and perception, while a leader is a person who is capable of influencing others. To lead others authentically means to rely on personal values, inner maturity, and the real self. The initiator of this leadership is Bill George, a professor at Harvard Business School, who states that leaders have to follow their own way instead of copying others. In other words, if we are not capable of creating an original and distinctive style of leadership based on personal values, knowledge, and well-established beliefs, no other copy will help to take the position of a leader, neither in an organization nor in life. Authentic leadership should be perceived as a naturally shown example for others. Suppose that by being sincere and responsible, leaders also spread these features in their organizations. By not showing fear of being vulnerable and by acknowledging their mistakes, leaders teach not to be weak and to be self-confident and strong. Whatever happens and no matter how we feel, we are able to maintain our strong position and not to break. By being respectful and equal in communication with colleagues, by sharing and including others, leaders show what kind of communication they expect from everyone else. Professor Bill George is convinced that every person is a potential leader. Everyone can become an authentic leader in their field if they learn to express themselves as individuals, know their own nature, and learn how to follow it, instead of acting as puppets by copying others. Although it is not emphasized what qualities are the most important for an authentic leader, a few usual tendencies of behavior can be named: desire to achieve important objectives which are based on their personal values, morals, and ethics; development of self-esteem and living in the moment; ability to keep sustainable, meaningful, and long-term relationships with people; relationships are identified as partnerships without any arrogance; personal discipline and order are maintained. Authentic leaders know their own pace, are stable, and do not give in to impulsive reactions, bad habits, or the daily rush; they know their strengths and weaknesses, systemically analyze themselves, devote time to self-knowledge, personal development, and growth. In addition, they create a psychologically safe environment by developing their followers. An authentic leader motivates team members of an organization to improve, experiment, and express themselves, and encourages them to learn from mistakes, be open to changes and share responsibilities. He might temporarily transfer leadership to a member who is more competent in a given area and at the same time become a part of it; leaders can take care of their self-image because of their self-esteem and respect for others; however, it is not an essential motivation (Biddle and Hamermesh 1998).

*The psychodynamic approach to leadership.* Psychodynamic theories provide several different approaches to leadership. These theories also are the personality of a leader. In psychodynamic theories, a personality means a consistent structure of thinking, feeling, and reacting to an environment. This theory emphasizes personality types and provides data which show that certain personality types are more suited for specific situations and duties of a leader (Diamond and Allcorn 2009). The premise of psychodynamic theories is that it is impossible to change personal characteristics that are deeply established. The most important thing is to acknowledge the traits of people together with their various oddities and to understand how these traits affect others as well as to recognize the characteristics and unique traits of followers. These leadership theories are based on the psychoanalysis of Sigmund Freud. Freud tried to understand patients and help those who had problems that could not be handled by usual treatment methods. Originally, he treated patients suffering from hysterical paralysis with hypnosis. Later the scientist noticed that hypnosis is unnecessary. In order for patients to recover, it was enough to encourage them to talk about their past. In this way, Freud created what we now call talk therapy. Freud had many followers. One of the most famous is Carl Gustav Jung, who eventually created his own theory of psychology. Today, Jung's psychology is widely recognized, while classical psychoanalysis has received less and less attention in recent years. However, the theory of psychodynamic leadership was based on works of both Freud and Jung. The most famous supporter of the psychodynamic approach to leadership was Abraham Zeleznik, a Professor of Management at Harvard Business School. One of the branches of the psychodynamic theory is called psycho-history, which attempts to explain the behavior of such historical personalities as Abraham Lincoln or Adolf Hitler. In these studies, historical data and biographical facts about leaders are analyzed. Various psychodynamic approaches to leadership are based on the same ideas. One of them is the concept of ego state, created by Eric Berne, which is also an element of a broader method known as transactional analysis.

*Transactional leadership* is based on a rational understanding of tasks and situations. In such management instructions of what subordinates have to do are common, their role and tasks are explained, their needs are identified, and conditions of fulfilling those needs after tasks are finished are named. When assigning tasks, a transactional leader does not take resources into account and does not expect followers to be open to changes, as they have to operate according to the indicated criteria. The manager identifies the needs of his subordinates, encourages them to gain self-confidence in order to achieve personal and organizational objectives, monitors improvement of employees, adjusts their activities, financially motivates them for completing tasks, or applies certain sanctions for not completing the assigned tasks or completing them to a poor quality. Peculiarities of transactional leadership: motivation by merits and management by taking exceptions into account, when subordinates are monitored and a reaction to their mistakes exists (Jung 2001). Such leadership is based on mutual benefit.

In the *transformational leadership* theory, unlike in transactional leadership, the irrationality of operational situations is acknowledged, and expectations of subordinates are rather changed (transformed) than satisfied, as followers are inspired to achieve important results and to improve leadership skills. This theory also highlights the emotional and charismatic elements of leadership, intrinsic motivation and education, emotions, values, and development of followers. Theory of transformational leadership provides means of general thinking about leadership and emphasizes ideals. Usually, the transformational leader has a clear and attractive vision for the followers or, more importantly, this vision is developed together with the members of the organization. The term pseudo-transformational leadership is used to describe a negative transformation (Samad 2012). Transformational leadership is very popular because of its attractiveness, easy perception, attention to the needs of other people, their improvement due to higher moral norms, greater responsibility of followers, effectiveness, followers' satisfaction in activities, and positive motivation to act and achieve better results.

All the transactional leader has to do is to change the former state, as he has more of a one-way impact; in charismatic leadership, success of activities and influence of a leader depend on important elements, such as personal skills, attractiveness, and appeal, whereas, in transformational leadership, changes are encouraged, new things are implemented, people achieve more than they thought they were capable of, and general welfare becomes more important than personal welfare, although the sense of activity that stimulates action is discovered at the individual level. According to this theory, a manager has to expand the needs of his followers instead of explaining or directing them. The leader has to cooperate and work with followers better than it is expected and motivate them to rise above their personal objectives by changing those needs.

The transformational leader is distinguished by attractiveness, individual attention to every follower, and promotion of creative activities. Such a leader values shared leadership, when leadership skills of followers are developed, the focus is on new leaders, and leadership is shared with followers if they are capable of becoming leaders themselves. By showing confidence in themselves and others, transformational leaders empower followers and protect them during the transformation period, and they create a culture of communication and tolerance for diversity by encouraging people to try new things and seek higher objectives. The leader might be both transactional and transformational. Other various forms of leadership might as well be applied in the combination.

Transactional leadership can be a part of transformational leadership, especially if it is effective. Transformational leadership is a process that changes people, and its outcome is measured by the level of followers' satisfaction in their activities. Various authors identify different main features of transformational leadership: *idealized influence*, *inspirational motivation*, *intellectual stimulation*, and *individual consideration*.

Scholars identify such strategies of transformational organizations: a clear vision of the organization's activities; transformational leaders are social architects of their organizations; confidence is created by transformational leaders; creative self-expression and self-esteem (Samad 2012).

Therefore, having a vision provides people with power, especially when that vision is based on the needs of the organization, as the mobilization of people allows the creation of a new identity of the organization, confidence is related to safety, and the emphasis on advantages, self-esteem, and education stimulate a positive reaction from the followers. The most widespread model of transformational leadership consists of five elements: demonstration of an example, instilling of a general vision, challenge of the process, opportunity to allow others to act, and encouragement.

The transformational leadership theory has received a lot of criticism because of its elitism, anti-democratic features, and exaltation of heroism.

*Transcendental leadership* is based on a relationship of personal influence. In personal relationships, employees are motivated not only by financial rewards and interesting work; they also have a personal commitment to the leader to conduct a mission that is valuable for everyone. The influence of a transcendental leader is even greater than the influence of a transformational leader, as such a leader can influence people not only by rewards, punishments, or interesting challenges; he can also question their perception of how people should do their jobs by feeling a sense of their mission. A transcendental leader is committed to a project of rich content and makes his subordinates understand how their work contributes to the implementation of that particular project. Examples are given in order to reinforce the leader's credibility among subordinates. Finally, such a leader shows a strong sense of necessity and encourages his subordinates to make leadership commitments in order to set hard and ambitious objectives for themselves, by serving the mission of the whole company.

A transcendental leader does not consider leadership as a core activity; he puts a lot of effort into spreading leadership throughout the whole organization. He is a leader that, first and foremost, develops other leaders. It is done by inspiring every subordinate with a sense of the mission, according to a specific level of their responsibilities. The resulting sense of ownership is deeper than the empowerment provided by transformational leaders.

A transcendental leader considers his work as a service to subordinates in order to help them achieve the mission of their level. Essentially, a transcendental leader serves the mission. For this reason, such leaders are more distant from their own opinions and even from their own work if that is required in order to achieve the mission. As a leader of leaders, he expects his subordinates to take on more responsibilities and he shares the achieved success with his employees instead of crediting himself for everything. One could say that a transcendental leader is not only more ambitious but also more modest than a transformational leader.

Huge global changes and continuous shifts motivate the search for alternative forms of organizational activities as well as for different methods of leadership; in other words, to find effective management solutions in order for the organization to prosper and maintain sustainability. Sustainable leadership is shared; however, not every leadership that is shared is the same as sustainable leadership. Shared leadership is the opposite of vertical, hierarchical, and formal leadership. It is creative and informal leadership. Shared leadership means sharing roles and methods of activities that can be separated, shared, modified, and followed consistently or simultaneously. Shared leadership occurs not only in planning; it is also expressed through roles, structure, and internal organization. Even though sometimes scholars separate *distributed leadership* and *shared leadership* by distinguishing specific subtle differences related to the roles of a leader and his followers, often these concepts are used synonymously, as in this book.

The essence of *distributed leadership* is shared activities and responsibilities initiated by a manager, which are distinguished by the will of a manager, use of power, and methods of activities. The essence of *shared leadership* is the responsibility shared by all members of the organization, distinguished by a higher degree of awareness, maturity, and professionalism among all members of the organization (Ross, Rix, and Gold 2005a, 2005b).

With implemented shared leadership, the method of management changes, since an individual promoted to the position of a leader does not have to perform functions of the leader. Scholars emphasize that, in the complex and dynamic world, leadership cannot be the responsibility of a handful of individuals. Hence, it is worth looking at the functions of the leader as functions shared by all members of the organization. However, shared leadership should not be mistaken with the delegation, i.e., when due to the increased amounts of work or the intensification of delegated functions of leadership, it is sought to share the hardest activities. In this case, it would be more appropriate to call it "the sharing of suffering." Shared leadership is not only a delegation of leadership to others, as it also includes aspects of power, authority, legitimacy, and politics.

Shared leadership is a fundamental and natural phenomenon of a modern organization. Scholars believe that models of leadership with a usual hierarchy of organizations and their leaders are transformed into networks where leadership is shared, dynamic, and distinguished by coordinated actions and cooperation. The form of leadership, independent of hierarchical structures, is related to skills, competencies, and values of the staff as well as being clearly focused on the education of all members of the organization and efforts to improve their skills.

In scientific literature, the idea of shared leadership is related to *sharing*, *collaboration*, *democracy*, and *participation* (Ross et al. 2005a, 2005b).

Shared leadership is not an area of activities of one person, as it is not restricted by any structural or organizational constraints. It is a dynamic result of many interactions, which are based on interrelationships in an organization where social context is an integral part of leadership. Shared leadership is now becoming the dominant model in both business and the public sector. Its success depends on the ability to create and manage groups of individuals, to disassociate from the relation between the leader and his follower, and to focus on interactions based on trust and consensus when completing the essential tasks. Nevertheless, shared leadership is criticized for its lack of clarity: who initiates the sharing, with whom, what is shared, and how it is shared. Different authors emphasize that shared leadership is not essentially a self-contained good thing since questions arise about the development and changes of the organization.

*Servant leadership* highlights the moral principle that people ignore formal institutional authorities and deliberately follow the one that is determined to serve others. It is a conscious and underlying service to others, a commitment and provision to lead others without raising oneself above them as a leader (Buchen 1998).

After the economic crisis, just like after all economic hardships, people tend to come back to a pattern of guidance and control, specific for the age of industry, and to spread the idea of moral authority or moral power acquirable through serving others. Even though this model, also called the "carrot and stick" model, can maintain the existence of an organization, it does not provide the best results. Servant leadership is a guarantee of implementation of changes. With this type of leadership initiatives are supported, the organization becomes emotionally stable, the microclimate of the organization is improved, and the community is brought together. Servant leadership is characterized by listening, empathy, consolation, compassion, understanding, persuasion, conceptualization, insight, management, and commitment to followers and community.

The theory of servant leadership focuses on the growth of followers as individuals, their inclusion in the development of the organization's activities and implementation of changes. By serving other people, leaders show their real power that has a real impact on others. This leadership is ethical, insightful, orientated to cooperation, making changes in the hierarchy pyramid, moving from vertical to horizontal (Russel and Stone 2002).

Scholars provide descriptions of five stages of managers (skilled individualists, active team members, competent managers, influencing leaders, and managers of the fifth stage) and emphasize the importance of managers of the fifth stage. These are modest and professional managers, who create long-term success. Their activities are distinguished by good management, mobilization of the community, and orientation to services. They are capable of creating trust-based relationships, empowering their followers, and increasing their self-consciousness. Such leaders serve themselves and direct their followers to serve others as well, and effectively use collective wisdom, knowledge, and experience. Usually, these leaders have strong and intelligent personalities, possess very good communication skills, and listen to others, as well as being understanding and compassionate. Such leaders prioritize the interests of others.

*Emotional or resonant leadership* is based on personal influence instead of formal power; for example, it depends on the emotional inclusion of members of the organization into the realization of common objectives (Waclawski 1998). It is based on the emotional intellect of the leader. In the broadest sense, emotional intelligence can be described as skills, more or less related to personal characteristics, that help process emotional information—for example, recognize and understand emotions and their meaning, assimilate and manage them and, on this basis, reason and manage members of the collective. The way the manager understands emotions is expressed in the understanding of the emotions of other team members. Correctly understood and expressed emotions help to motivate and inspire subordinates: it is much easier for the manager to understand subordinates' motives and things that motivate them to perform better, which is the reason why the manager adapts his actions to the values and motives of his subordinates. In addition, the manager's tolerance, respect for other people's feelings, and sincere relationships will also help subordinates feel safer, and to communicate more openly among themselves and with the manager. In this case, unnecessary tension is eliminated, and favorable conditions are created which allow employees to focus and work in a more productive way. By hiding feelings, especially negative ones, the working time of subordinates is being wasted, creativity is reducing, and the fear of risk is increasing. Without expressing their feelings, employees can evaluate a particular situation inadequately and, due to this, wrong decisions can be made. This might also stimulate an increased employee turnover, a waste of time for things not related to work, ineffective communication in the organization, various misunderstandings, and loss of work efficiency. After satisfying their emotional needs, subordinates feel more productive, motivated, patient, and creative. Therefore, it is especially important for a good manager to learn to recognize emotions and to manage them. The development of emotional intellect helps to do that.

*Visionary leadership.* At present, efficient leaders recognize that leadership processes, stages, and methods are implemented through people. Current dynamic changes in business and its environment dictate that leaders have to include team members into the implementation of an inspiring vision for the future. From the researchers' perspective, effective leaders should inspire and motivate their team members in order for them to become enthusiastic followers of the leader and to commit to his vision (Ensley, Hmieleski, and Pearce 2006). Visionary leaders are creators of a new organization, as they work with a clear vision and insight in mind as well as show courage when tackling challenges of the future. They are capable of challenging members of their team and bringing together the best people for the implementation of their vision. Such leaders are social innovators and agents of change, who clearly see the vision of the wanted future and think strategically. Usually, such leaders are looking for unconventional decisions and new paths in order to achieve a vision and they find success since they have a clear vision of the future. This type of leadership is very important when changes are initiated, and new ways for development are found, and it can also be applied in the transformation of a more sustainable organization (Dhammika 2016).

*Value-based leadership* is a type of leadership in which leaders, relying on values, effectively communicate values of the organization and direct members on how to behave in order to implement the mission of the organization. These leaders are capable of relating organizational values to the personal values of members and in such a way ensure the commitment of employees to the organization and its mission. Value-based leaders rely on fundamental values that reveal the strengths and character of the organization. Since the underlying values reflect the essence of the organization, they are maintained even in the conditions of the changing market and globalization. In order for the employees to sincerely believe in the values of the organization, leaders and their teams have to be an example and present those values to all employees. The efficiency of values depends on how strongly they are rooted in the whole organization. Value-based leadership is related to the management of missions. It can also be successfully employed when transforming the organization into a sustainable one because the values of a sustainable organization are important to society and to every individual (Bethel 2004).

*Mission-driven leadership* was created by Pablo Cordona and Carlos Rey in 2008. It is a special form of transcendental leadership. In real life, there have actually been managers, so-called transcendental leaders, who create other leaders. The majority of them are widely researched, admired, and considered examples. Usually, these are people with unique special characteristics, capable of bringing together a team of leaders. They have comprehensively implemented principles and values that helped them to ensure things that current companies are trying to achieve: employees who are committed to a trustworthy and necessary mission of rich content. And yet these are very exceptional cases with exceptional people and exceptional achievements.

Still, transcendental leadership can be achieved at all levels if the situation is appropriate. Then, a mission-driven management system can be implemented since it creates a special form of transcendental leadership called a mission-driven leadership (MDL). According to MDL, to manage people means to turn them into leaders, who will take on a mission on their own level and will implement it perfectly. Unlike transformational leadership, MDL can be developed at all levels of the organization. According to MDL, managing departments, subdivisions, and even people is not that important. The most important thing in MDL is that every manager at every level manages a mission and that makes them leaders. It is not due to the fact that the mission-driven leader has particular traits or special attractiveness. MDL is directly related to a mission and values which are above the personality of an individual leader. First of all, the leader is a person who is directly committed to the mission in which his subordinates participate. The MDL management means are the driving force and the leader is the mediator of the cultural change in the company. In other words, cultural change in the organization does not occur automatically, just because the company uses specific means. It is a process of teaching and learning, during which managers and their subordinates acquire new knowledge, attitudes and models of behavior, while everyone takes responsibility for managing missions at a certain level. MDL has been successfully implemented in many companies while experimenting with a model in which the manager can become a mission-driven leader. This model consists of three main aspects: commitment, cooperation, and change.

*Spiritual leadership* is a relationship-oriented leadership based on spiritual values (spiritual intelligence) (Barrett 2003). Spiritual leadership is used to achieve objectives of the community and organization by helping members of the community to discover their calling and to develop it; in such a way, everyone, based on the personality and talents, is welcome to participate in the implementation of the objectives of the community and organization. Since the modern world is very dynamic, changes in business show signs of a need for a leader who is distinguished by new wisdom (spiritual intelligence). Leaders of the new age have to take this into account and actively improve themselves as well as help their followers become their own personalities too. The more a leader understands the environment, customs, values, and traditions of the organization, the greater his impact is. The theory of spiritual leadership is based on a model of inner motivation, which includes vision, hope/faith, altruistic love, theories of workplace spirituality, and survival of spirituality. Spiritual leadership theory can be partially considered as a response to the need for a more holistic leadership that helps to include the four main areas which describe the essence of human existence in the workplace: body (physical), mind (logical or rational thoughts), heart (emotions and feelings), and spirit. Previous theories of leadership focused on different degrees of physical, mental, and emotional dimensions of human communication within an organization; however, spirituality was not considered a separate type of wisdom (intelligence). Therefore, spiritual leadership is a value-based behavior and activities that unite and motivate all members of the community and organization, since this leadership provides a feeling of spiritual existence through the calling and membership. Importance and relevance of such leadership have increased dramatically. Often people care more for their physiological and emotional existence. Meanwhile, staying true to oneself becomes more and more valuable. It is more of a search for oneself than an intellectual choice. Everyone can improve by having everything that is needed to replenish and grow. The more improvement will be sought by a mature personality of high culture, the more this person will follow the humanistic values that are valuable, give meaning to life as well as spread progress in the human reality. Self-creation takes a very important role here. Self-creation is great human power. A spiritual leader is a person who, based on previous experiences and world view, finds and provides a functionally correct solution to the action with regard to social gestalt (part of which he is himself) as well as receives a moral reward and spiritual expression for it. A broad spectrum of knowledge is not enough for a spiritual leader. Such leaders have to perceive it and integrate it into a particular level of consciousness through their understanding and visions. A conscious leader has to comprehend complex knowledge through all forms of wisdom: intellectual, emotional, and spiritual. But first, it is necessary to know that such forms of wisdom (intelligence) exist, to know their origin, and the possibility to develop and use them.

*Moral leadership* is close to spiritual leadership; however, this concept has more distinguished features. Alexandre Havard is the founder of moral leadership. According to this scholar, moral leadership applies the wisdom of ancient European philosophy to a modern organization. It goes beyond the usual sense of leadership that includes charisma, management style, and employment of various manipulation techniques. Moral leadership encourages an overview of the beliefs of leadership and the criteria of the professional and personal success in the light of classical virtues. Moral leadership helps everyone achieve personal excellence by developing the virtues of spirituality, wisdom, justice, courage, and restraint. Moral leadership is not a skill nor a method, but a way of life. It is character, virtue, and behavior. It shows what kind of impact virtues have on effective decision-making, successful strategic planning, the creation of consensus, efficient communication, strengthening of trust and loyalty, motivation, development of a manager's personality, and creation of authentic organization's culture. Moral leadership helps to better understand classical virtues and to practice them. It allows everyone to recognize their weaknesses and strengths in the area of leadership and encourages professional as well as personal improvement. Since the principle of ethics is very important in leadership, this particular type allows avoiding unethical expressions of leadership—for example, various activities of dictators.

*Results-based leadership* describes a leader who prioritizes orientation to achievements. Such a leader is depicted as one who sets maximum objectives, motivates followers to put in more effort, and provides more confidence in the capability of followers to perform their tasks.

Globalization and the progress of science and technologies constantly bring changes to the economy, society, and culture, as well as processes of organization management. The ability to change and accept innovations, creativity, fast reaction to changes, and various models of activities are incompatible with hierarchical leadership. Efficient leadership is particularly needed when fast and constant changes occur as well as when the needs and expectations of people grow. Thomas L. Friedman states that in "the flat world" leadership has to embody creativity, flexibility, mobility, and ingenuity despite the structure or limits of an organization.

In the *latest theories of leadership*, New Millennium or New Paradigm leaderships are described by emphasizing the importance of holism, intuition, and creativity, as well as the systemic perception of the world in the realization of leadership powers. More examples of effective leadership appear that are checked in practice, based on analysis, and recommended by the field of management science.

The new concepts of leadership show a change in the relationships between the individual and the organization as well as the impact of innovative technologies and networking. Nevertheless, leadership is still desirable; however, not as the power of the chosen ones but as a universally needed competence. The key challenge is to find and develop future leaders.

The UN, by highlighting that people overstep limits of using resources, which can no longer be restored, and that the foundation of the ecosystem and biological variety are being destroyed, has proclaimed a Decade of Education for Sustainable Development (2005–2014). Of course, it is important to ensure sustainability not only in the environment but also in other areas of life. Sustainable leadership has an important role to play here.

*Sustainable leadership* focuses on aspects of continuity, the efficiency of changes, long-term motivation, and satisfaction in activities. Charismatic leaders can inspire their followers and together achieve unbelievable results; however, after such leaders have resigned or "burnt out," high standards are not always maintained or constant improvement is not always achieved since followers do not manage to continue their activities successfully. Scientists emphasize the energy constraints of leaders, the excess of the initiative, and the chaos of change. Lately, the frequent change of managers in organizations has become quite popular; however, as research carried out in Canada, Great Britain, and other countries show, it does not guarantee the progress or improvement of an organization (Doppelt 2010). Continuous training of managers is necessary for the efficient change of them. Sharing responsibility, practice, ideas, and duties is crucial for sustainable leadership.

Sustainable leadership is a long-term activity, designed to achieve an organization's objectives, and is based on values that help improve organizational activities, overcome occurring internal or external threats, and reduce emerging risks.

Individuals, managing in a sustainable manner, perform more successfully and indicate how such types of organization operate:

- the objective is valued more than profit;
- long-term objectives are maintained through changes;
- a slow start and steady progress;
- no dependence on one leader with a vision;
- new leaders are developed in an organization instead of hiring them;
- learning through experimentation.

In sustainable organizations, a strong moral foundation is implemented, and important principles are set: the responsibility of managers, transparency of activities, community, honesty, appropriate behavior with employees, sustainability, diversity, and humanness. Authors indicate these elements of sustainable leadership: moral values, influence, creativity, teamwork, spiritual environment, opportunities for learning and development, policy and development of individuals, learning according to the context and system, leadership in all stages, and improvement of the development (Flowers 2008). Some authors highlight the aspect of attention, hope, and compassion and state that sustainable leadership is meaningful at the individual, team, organizational, and community levels. Sustainable leadership is divided; however, not all divided leaderships can be compared to sustainable leadership. The autocratic division does not result in sustainability because sustainability is associated with sincere responsibility.

All discussed forms of leadership highlight one or more key elements of leadership (power of personality, interaction with followers, vision, values, and orientation to results); however, all these types of leadership can be combined with the concept of sustainable leadership, based on the concept of sustainable development and sustainable organization, since the most important aspect of sustainability is the continuous preservation of economic, social, and environmental values. Together all elements of leadership correspond to the most important economic (orientation to results), social (interaction with followers and orientation to people), and environmental (values, vision, and mission) aspects, as well as allow the leader to transform the organization into a sustainable one by using power of personality, sustainable management, and in this way also changing its culture. It is important to note that sustainable leadership has two concepts. One concept focuses on the longevity of leadership and its development, while the other is concentrated on the compatibility of social, environmental, and economic elements (compatibility of rational, emotional, spiritual, and physical intelligences) in the efficient activities of the leader.

The concept of sustainable leadership will be formulated in the following chapter of the book; nevertheless, in this part of the book, dedicated to the meaning of leadership and its role, it is necessary to discuss the influence of leadership on the psychological resilience of employees, as the quality of the workplace, satisfaction with work, and quality of life can also be ensured by the tools of leadership.

## 3.5 APPEARANCE AND IMAGE OF THE LEADER

Nowadays image creation and management issues are highly relevant to the specialists of public relations, marketing, and various areas of management, as well as organizations, together with their managers and employees. The external image of a person can consist of many different factors: clothes, attractiveness, voice tone, vocabulary, facial expressions (mimic), eye contact, gestures, and social behavior, i.e., knowledge and application of etiquette, etc. An image of a person is a powerful tool when seeking a career and good management results at work and in performing other important social functions.

The image is primarily the result of the personality perception of an individual, a notion in the mind of the perceiving person (Bogdanov and Zazykin 2003). The image may also be defined as beliefs about the perceived person and feelings regarding him or her. In addition, it might just be what comes to our mind when we think of the individual. By looking at the definitions, we can see that the image is subjective. The word image is not accidentally related to the word imagination; however, it is not merely the result of imagination. It is the reflection of someone's personality, which is formed in another person's or our own mind. Therefore, some authors even claim that it is more appropriate to use this word in the plural since, in the eyes of different people, the same person can have a very different image. We also have an image in our own eyes, which is a self-image (Mamedaitytė 2003). In addition, there is the obtained (existing) image, how others perceive us; the image that is being sought (created) is the way we want to be perceived; and the perceived image is the way, in our opinion, others perceive us.

As it was mentioned before, leaders are distinguished by their tendency to dominate, non-traditional behavior, desire to influence, ability to inspire others, high self-confidence, high self-esteem, and great authority. The leader always looks good, has very high confidence, shows an example to others, is competent, communicative, attractive, has an excellent image and enviable reputation. It is a person you want to resemble or at least emulate. Such a leader has a tremendous impact on his followers, who admire and sincerely trust him. Thus, an excellent appearance and appropriate behavior are important parts of a leader's good image. Messy and dirty clothes, poor taste, as well as appearance and behavior which do not correspond to the situation and position are not compatible with the image of the leader that is very important to his followers.

Economists from the University of Wisconsin Joseph T. Halford and Scott Hsu (2014) have found out that there is a direct link between the attractiveness of the leader-manager and the company's share price. Specialists who carried out the study, "Beauty Is Wealth: CEO Appearance and Shareholder Value," analyzed 677 managers from companies belonging to the S&P 500 index, by focusing on their facial features. It has been revealed that good-looking managers earn much more due to the so-called "beauty premiums." In addition, during the first working days, these leaders record a higher return on the shares of the companies they manage, when they appear on television; the price of these shares noticeably rises.

The results show that physically more attractive managers-leaders earn more for one specific reason: they create value for shareholders, as then it is easier for them to negotiate, and they are more visible in public life. The study also revealed that the attractiveness of a manager-leader also brings more benefits when announcing acquisitions. The gathered evidence suggests that attractive managers gain more benefits from different transactions for their managed companies. This fact justifies the hypothesis that more attractive managers, who have an obvious advantage in negotiations, create greater value for the shareholders. The authors of the study claim that Marissa Mayer, the 38-year-old president and chief executive officer of Yahoo!, is an excellent example of an attractive manager-leader who has a positive impact on the company's shares. In the index of facial attractiveness, she scored 8.45 points (out of 10) and ranked among the top 5 percent of all analyzed managers. Since July 2012, when M. Mayer, called the goddess of Silicon Valley, began to lead Yahoo!, the company's shares jumped to the top. "Of course, we don't mean that all the increase in stock price is from her appearance. We just find that there might be some positive correlation between the two," say the authors of the study. J. T. Halford and Hung Chia Hsu emphasize that the physical attractiveness of a manager-leader is not the only factor that is important when hiring a new director. Other features must also be taken into account. When looking for a good manager-leader, appearance is not the only important feature. However, for companies that are focused on negotiations and representation, more attention should be paid to the good image of the manager-leader and his physical attractiveness.

Professor J. Biddle of the University of Michigan and his colleague from the University of Texas, D. Hamermesh, focused on the appearance and its influence on the image of the leader, the manager's career, and the success of the business. Their extensive research shows that attractive people, who are well-dressed and who pay great attention to their looks, are treated more favorably than those whose appearance is less attractive (Hamermesh and Biddle 1994). This easy form of discrimination exists everywhere: from schools and universities to companies and other organizations. One of the many examples could be a story, from 1994, in Texas. One woman applied for a bus driver's position at the local public transport company. The employer did not hire her solely because she was overweight. Another similar story took place in 2005 in New Jersey. A woman worked as a cocktail waitress at one of the casinos in Borgata hotel. She had health problems, influenced by thyroid activity; for this reason, she slightly gained weight and was fired. In 2001, a responsible and qualified aerobics instructor was dismissed for being overweight in California. The reason for the dismissal which was given to the employee was that his weight did not correspond to the aerobics federation standard. These are only a few cases which raise the question of the importance of a person's appearance at work. In the United States, a study was carried out which showed that 62 percent of women and 42 percent of men who were overweight had suffered direct discrimination at the workplace from the employer's side or were not employed due to them being overweight. The results of this study should be looked at more widely. In the study, the causes of discrimination at their workplace, identified by workers, were related to being overweight or, in other words, to the physical appearance of people. Another study, also carried out in the US, has shown that on average about 16 percent of people have been discriminated against at work because of their unattractive appearance. Moreover, unattractive individuals are less popular in society, and they are seen as less intelligent or smart.

A study carried out at Cornell Law School showed similar tendencies. There was a simulated court hearing with two types of accused: attractive and unattractive students. The results of the study showed that physically attractive respondents on average received 22 months' shorter sentences than unattractive respondents. The image is directly related to the natural, innate physical appearance of a person as well as his clothes, manners, and verbal and non-verbal communication. Attractive appearance and physical beauty of a human being is a relatively complex issue. What looks beautiful to one person may look the opposite to another. However, in society, there are certain beauty standards; this beauty standard of a Western European man could be given as an example: tall, slim, fit, and muscular. Nevertheless, human beauty standards are not eternal; they change constantly. Moreover, in different cultures, different perceptions of beauty standards dominate.

In a study carried out by J. Stossel, four persons participated: two men and two women. One participant of each gender had an attractive appearance, while the other one looked less attractive than an average person. All four participants had to attend a job interview. They were taught the same rules of conversation and behavior manners. Their only difference was their natural physical appearance. The results of the study showed that both attractive-looking people were recruited, whereas the less attractive woman and man were not employed (Stossel 2006). In addition, communication with the attractive study participants was more pleasant and polite than with participants whose physical appearance was below average. The conclusions of the study carried out by J. Stossel confirm that the appearance of a human being and his created image really impress people. A vivid demonstration of that is the US Presidential Debate between J. F. Kennedy and R. Nixon. Americans who listened to the debate on the radio were sure that the debate was won by R. Nixon; however, those citizens who watched the debate on their television screens attributed the victory to J. F. Kennedy. During the debate, J. F. Kennedy looked strong and energetic, while R. Nixon seemed tired and pessimistic. People who watched the debate attributed their victory to the person who managed to create a positive image and looked good. The results of the study, carried out by J. Stossel, were confirmed in 1985. Forsythe, Drake, and Cox (1985) conducted research with respondents who were looking for a job and participated in an interview with a potential employer. The results confirmed that physically attractive people who wore proper and tidy clothes got a greater number of points in the interview with the employer than those who were not as good looking and wore casual clothes that did not match. Another interesting study has helped to determine that, in most cases, a simpler and less rigorous interview process is applied to those people that look more attractive during a job interview. Often, the conversation is shorter, more enjoyable, and fewer complex questions are asked in comparison with the individuals with poorer looks.

Quinn (1978) conducted a study involving several US companies. Respondents were asked to provide information about employees who had applied for a job and had an attractive appearance. They were also asked about the wages earned by the same employees. The study revealed that attractive employees earned more than not so good-looking people, though they worked in the same positions. These results are confirmed by Hamermesh and his colleagues who, in 1994, carried out a study in

Shanghai (Hamermesh and Biddle 1994). This study revealed that naturally beautiful women get higher wages (the physical beauty of men is less affected by this). Moreover, less attractive women who tend to invest in beauty products, such as clothing, cosmetics, etc., have only a slight impact (in a positive direction) on their wages. The study of D. S. Hamermesh and his colleague J. E. Biddle continued for years (Hamermesh 2011). In 1971 and 1977, 2,164 and 1,515 employees were interviewed in the US, respectively. In addition, in 1981, 3,415 employees were interviewed in Canada. In all three surveys, the interviewer assessed the physical appearance of each employee on a scale of 1 to 5 (Table 3.7).

The appearance of approximately half of the respondents was rated below the moderately attractive level and the remaining part above this level. Respondents were divided into three categories:

- those who are below the moderately attractive level;
- those who are at the moderately attractive level;
- those who are above the moderately attractive level.

The authors calculated the average wage earned by each category of respondents. The study has shown that those employees who belong to the moderately attractive level or are above it get higher salaries than those who are below the moderately attractive level (Hamermesh and Biddle 1994). In 2005, Hamermesh and Parker conducted another piece of research with students. Photos of all respondents who participated in the study were also evaluated on a five-point scale (Table 3.7). The results of the study showed that those students (regardless of their gender) who successfully graduated and were more attractive earned more than those who were less attractive.

Another interesting study was carried out by Harper (2000) in Great Britain. A group of several people, born in Great Britain on March 3, 1958, was formed. The study took quite a long time because every respondent was contacted five times, according to the following schedule:

- when the person reached the age of 7;
- when the person reached the age of 11;

**TABLE 3.7**
**Evaluation Criteria**

| Evaluation scale score | Appearance description |
|---|---|
| 1 | Indistinctive |
| 2 | Less than moderately attractive |
| 3 | Moderately attractive |
| 4 | Quite attractive |
| 5 | Attractive |

*Source:* created by authors.

- when the person reached the age of 16;
- when the person reached the age of 23;
- when the person reached the age of 33.

During the first two contacts with the respondents (when the respondents reached the ages of 7 and 11), their teachers were interviewed. The teachers were asked to describe the behavior, attitudes, character, and other social characteristics of the respondents. Furthermore, the teachers were asked to evaluate the physical appearance of children using a five-point system, according to the scores given in Table 3.7. In the following stages (when they turned 22 and 33 years old), those respondents who were evaluated as attractive were more successful at work than those whose appearance was rated as less attractive. In some sense, wages can be treated as the evaluation criteria of the employee.

Another study has helped to clarify the fact that those teachers who are physically more attractive receive a better assessment of their work as a teacher. In general terms, the results obtained by assessing university teachers confirmed that those departments with a higher number of physically attractive teachers were more favorably evaluated. Discrimination based on a person's appearance is a silent form of discrimination that certainly can occur as not receiving a promotion, lower wages, etc.

An interesting question arises: how much more does a physically attractive person earn? This question was examined by Zakas (2005), who carried out a study with Dutch participants, as well as Hamermesh, Meng, and Zhuang (2001), who conducted their studies in the US and Canada. The results of their studies have shown that a physically attractive person earns 12–13 percent more than colleagues who do not have such a distinctive appearance. Further research by various authors has shown that employees' premium for their attractiveness goes on throughout their career. Subsequent studies have even revealed that not only do good-looking employees start working with higher wages in new positions, but also their earnings grow faster in the long run. Biddle and Hamermesh (1998) conducted an interesting study with students who had graduated from one of the law colleges. Their appearance was assessed from a picture on the scale given in Table 3.7. The study revealed the following results:

- After five years, better-looking lawyers who graduated in 1970 earned more than the students who also graduated the same year. Other factors, such as grade averages, were equal in the subject group.
- Lawyers who worked in the private sector had a more attractive appearance than those lawyers who worked in the public sector.
- Better-looking lawyers reached higher career heights than those whose appearance was underestimated.

Following this study, the above-mentioned authors have taken interest in how the attractive appearance of people, in relation to the growth of income and different distribution of income, can impact even faster economic growth. Attractive-looking employees are often more productive than workers without such attractiveness. In addition, research has shown that customers tend to make a transaction and purchase a product or a service from a company employee who is physically attractive. In another study, respondents were asked: is the employee's physical appearance irrelevant to work? Eleven percent of respondents have said that appearance is very important, while 39 percent of respondents have said that appearance is somewhat important.

In 2005, in a US study (HRM Guide 2005), the appearance of employees in different companies was assessed. Appearance was rated according to the following criteria: physical build, appearance, and hairstyle. Then, a survey was carried out which involved 1,000 Americans. The respondents had to assess whether they felt discrimination at work because of their appearance. Thirty-nine percent of respondents said that the appearance of employees did not affect the relationship with the employer, while 33 percent of respondents answered that physically good-looking people have more job opportunities in their workplaces. The remaining respondents who were physically less attractive replied that employers who pay attention to the appearance of employees should be restricted by law. Summing up the results, it can be said that 61 percent of respondents felt that their appearance has an influence on their evaluation as employees. The same respondents were also asked to give reasons for discrimination in the workplace. Thirty-eight percent of respondents said they had suffered discrimination because of their appearance, 31 percent for weight, 14 percent for hairstyle, and the rest for other reasons. The first three mentioned reasons are directly related to the employee's appearance. Thus, the study shows that 83 percent of respondents who have been discriminated against at work have experienced it primarily because of their physical appearance.

Zakas' (2005) research, conducted in the Netherlands, also aimed to determine whether people's physical appearance could lead to discrimination in their workplace. The majority of the respondents of the study confirmed that, at their workplaces, co-workers with attractive looks have more advantages than workers who are not physically attractive. Another objective was to determine how important the first impression of a person's image is in a job interview. The vast majority of respondents confirmed that in the job interview good image and good looks, created by the applicant, play a significant role. But why? A joint study conducted by Massachusetts General Hospital, Harvard Medical School, and Massachusetts Institute of Technology revealed that certain chemical reactions happen in the human brain when looking at a physically attractive person (Kennedy, Makris, Herbert, Takahasni, and Caviness 2012). The participants of the study were a group of heterosexual males who looked at the preselected photos of physically attractive and unattractive men and women. During the investigation of participants' brain activity, it was revealed that, when looking at the image of an attractive person, those brain areas and functions that give a positive response to the requested or desired thing are activated.

According to research data, it can be stated that the physically attractive appearance of people influences their employability, earnings, and career advancement opportunities. A well-dressed, good-looking employee with an attractive appearance and good image usually gets a higher salary, better job results, and a faster rise up the career ladder, more easily becoming a manager-leader. Of course, just a great look is not enough. Taking into account the abundance of research results about the positive impact of employee appearance and image on the performance of organizations, companies are starting to invest in a specific area, i.e., the appearance of their employees. The benefits are mutual: for companies, their investment pays off quickly and generously due to satisfied customers; another important aspect is that employees that are satisfied with the attention and opportunities provided by the company show even better performance results.

In addition, scientists have proven that attractive-looking individuals have better social skills, they are more communicative, and thus by communicating with corporate customers, they develop a specific human capital that influences better customer relationships. Zakas (2005) analyzed 289 Dutch advertising agencies between 1984 and 1994. Six photos of the employees were collected from each company: of two men and one woman over 40 years old, as well as of two men and one woman up to the age of 40. Observers, selected according to the same criteria, evaluated the photos of the employees of all companies involved in the research. The scale from 1 to 5 is presented in Table 3.7.

The conducted research showed that, on average, the appearance of the advertising company employee was 2.8 points. This study also assessed the performance indicators of companies, such as sales volume and earnings, and average employee wages. It was found that, despite other indicators, the physical appearance of the employees influenced positive changes in the company's sales and profits. In addition, the appearance of employees has been linked to the company's own growth in the market. The study has shown that, if the physical appearance of an employee is above average, he or she earns a higher wage than the one whose appearance is below average.

According to research data, it can be reasonably assumed that the appearance of company employees influences even the image of the company itself and thus ensures the company's success in the market. The studies examined above show that an attractive employee influences the company's activities: sales volume and earned profit, as well as affects the competitiveness of companies, their performance, and growth results.

Summarizing the results of the abundant research discussed above, it can be stated that people's appearance as well as positive occupational and personal image have an impact on their career and ability to establish themselves in the organization, and to hold high and responsible positions; it also creates good conditions for establishing and maintaining the image of a manager-leader. Managers who have an attractive appearance, good looks, and a good sense of style seem competent and trustworthy, making it easier for them to gain a leader's position and trust, bring together a team of followers and lead it to the implementation of important organizational objectives and the organization's mission.

## 3.6 LEADERSHIP AND PSYCHOLOGICAL RESILIENCE OF EMPLOYEES

Currently, stress and fatigue are constant attendants of employees. Since organizations can increase operational efficiency by developing the psychological resilience of employees, much attention is paid to determine how a manager can promote resilience of employees.

Stress is a human state arising from various extreme effects, i.e., stressors. These are the protective reactions of the organism caused by harmful external or internal factors (Jones et al., 1988). There are many events and circumstances which may cause stress. However, how we react to stressors is influenced by our approach to the environmental requirements that affect us. An individual can respond to stress adequately by using his abilities, but if he does not adapt, then he does not feel well, and this dissatisfaction turns into certain negative processes in the organism. Hence, a person can avoid, reduce, or eliminate the damage caused by stress by choosing the appropriate response (by actually assessing the situation and his/her strength and competence to deal with the situation). In addition, once an organism has successfully overcome the effect of a stressor, it acquires a certain immunity, practices, gets used to it, and reacts to this stressor less in the future. At this point, the link between psychological resilience and stress is most prominent: resilience is the immunity that is being developed by successfully solving the challenges of unfavorable circumstances (Dewe 2002). Harmful factors (events or situations) may be divided into individual and organizational factors.

It is important to analyze sources of stress at work; therefore, we will continue to study stress-related variables in the work context. Individual variables that affect stress at work are the following: type of behavior, the degree of control, gender, social support, response to stress, etc. The studies show that employees with burnout syndrome believe that they cannot control the consequences. When the employee has a low degree of control, and the task requirements are high, there is a high likelihood that he will suffer from stress. It is also noted that women undergo certain stressors worse than men. For example, women are more likely to experience work–family conflicts than men. In terms of social support, studies have found that individuals who get more support experience a lower degree of emotional exhaustion and a higher degree of personal achievements.

On the one hand, harmful work factors are related to work content, work structure, management systems, work environment, and organizational conditions, but on the other hand, they are also linked to the competence and needs of the employee (Ahmed and Ramzan 2013). Causes of organizational stress: role ambiguity, overload or underload (for example, stress arises when a person's skills are underused), role conflict, pace of work (it is an unfavorable situation when a person has a low degree of control of the pace of work), shift work, task attributions (a person is stressed if there are few opportunities to communicate at work, and he is an extrovert), as well as relationships with co-workers, managers, and customers. Studies have shown that social support reduces the negative consequences of stress, while attentive leadership is associated with less stress at work. A decentralized organizational structure, where the power of decision-making is greater, causes less stress at work. A high organizational culture, when there is no negative and competitive tension among the employees, also reduces stress at the workplace, as it is noted that competition is a factor mostly related to stress at work. Feedback and career prospects also have a significant impact, as stress can be caused not only by non-promotion but also by too high promotion, a discrepancy of status. Stress due to reward is experienced when the reward is inadequate. The consequences of stress can be physiological (changes in heart rate and blood pressure), psychological (depression, burnout), cognitive (reduced concentration, impaired memory, distorted perception, etc.), and behavioral (absenteeism, employee turnover, bad habits). It has to be noted that the consequences of stress on people can be both positive and negative. When stress does not exceed a certain limit, it has a positive impact (encourages improvement, growth, more intensive actions, faster decision-making). If stress exceeds this limit, the impact becomes harmful (disturbs activities, causes the feeling of helplessness, drains mental energy). This optimal limit is individual for each person. It is claimed that the problem of resilience is primarily a problem of development of relevant personality traits. The way a person reacts to stressful, negative events and circumstances depends on his personal characteristics. Characteristics of an individual have an influence on whether he considers events and circumstances as stressful and what methods of coping he applies. In this context, frequently mentioned variables are endurance, sense of inner harmony, the outcome of inner control, sense of self-efficiency, optimism, and self-esteem. Not in vain, most of these characteristics are frequently used as synonyms for the term "psychological resilience."

Thus, people's ability to distance themselves from those environmental impacts that are contrary to their attitudes, habits, and motives, and the ability to adapt to the effects that are useful or inevitable, is one of the main signs of high psychological development associated with certain personality traits and personality in general. Therefore, in terms of psychological resilience, several levels of its expression are differentiated: the first one describes the expression of certain personality characteristics, while the second one is when this expression is identified as one of the criteria of personality development (Seibt, Spitzer, Blank, and Scheuch 2009). A mentally resilient person must be able to control his mental states and change them in a favorable direction, i.e., to manage self-regulation mechanisms (ability to control oneself purposefully). This process is determined by the peculiarities of the personality's mental activity, personal characteristics, style of action, and specific circumstances.

It is noted that psychologically resilient individuals have certain qualities that distinguish them from vulnerable individuals. Thus, the qualities describing the expression of psychological resilience are the following: social competence (empirical approach, ways of communication, sense of humour; these qualities cause positive reactions from others); effective problem-solving skills (the ability to think abstractly and reflexively); adaptive solutions (the ability to quickly generate several possible solutions as well as the ability to modify them when they appear to be ineffective); autonomy (the ability to control the environment, to withdraw from the dysfunctional environment, independence); orientation to objectives (formulation and achievement of objectives); planning (development of strategies to achieve objectives); internal sense of control (a feeling that a person controls his own destiny); ingenuity (clever access to information, real sources); social support (strong tendencies in seeking the support of others; possibility to get help, support from family, friends, co-workers as well as willingness to support others); positive approach to the future (positivity despite unfavorable obstacles; failures are seen as a natural result of the changing world); self-esteem (self-confidence, evaluation of one's strengths and recognition of one's limitations); evaluation of changes (changes are seen as an opportunity for improvement, but not as a threat or a crisis); self-discipline (the ability to set priorities and to multitask); flexibility (the use of internal and external resources to develop creative and flexible response strategies to changes); activeness (in situations of uncertainty and vagueness people risk, rather than seek comfort); correspondence (the correspondence between values and beliefs of an individual is important); sense of well-being (a common sense of mental well-being that helps to overcome difficulties). All these characteristics are significant for a leader. Thus, psychological resilience could be identified as an essential feature of a leader.

Many authors associate psychological resilience with constructive thinking, hope, ingenuity, optimism, and self-efficiency. Studies confirm that positive emotions encourage psychological resilience during negative events. Very often the lack of mental resilience is associated with the lack of such characteristics as good self-assessment, the ability to express and manage emotions, and the ability to constructively overcome the difficulties that arise.

The results of studies on the positive consequences of traumatic events reveal that psychologically resilient individuals see their environment as promising (optimism) and other people as non-harmful (trust). These attitudes of the resilient individual are not unconditional because, in such a case, it would be a sign of naivety (Seibt et al. 2009). Psychologically resilient individuals perceive failures and stressful events as a normal and meaningful thing. These people believe that life can be influenced (internal locus of control) and they see themselves as capable of doing so (self-efficiency). Stressful events are seen as challenges (endurance). Resilient individuals are emotionally stable, and they have no tendency to experience negative emotions. Some of the mentioned characteristics are inherent, and some of them develop in the course of life. Psychologically resilient personality traits act as protective factors. These traits are also related to productivity, involvement, and progress. As these processes are very important for business success, managers increasingly focus on identification and assessment of resilient personality traits during recruitment. On the other hand, since psychological resilience is an unstable construct, which responds to changes, it is necessary to create an environment that promotes and supports resilience and to form an appropriate organizational culture. It is important for every organization to have psychologically resilient leaders and to seek that psychologically resilient employees would work there. The conducted research confirms the fact that psychological resilience is a construct that benefits both the individual and the organization. It is found that resilient employees have a high commitment to the organization, improved job satisfaction, and a stronger sense of justice (Shah et al. 2012). Such an employee is less influenced by stressors related to his role (role conflict, work overload) and tension at work. A psychologically resilient member of the organization is less likely to leave his position. In addition, such employees are more likely to plan and engage in introspection and self-assessment, which can improve and achieve better results in self-expression and work.

Employees with low resilience are characterized by frequent morbidity and incapacity for work, absenteeism, more frequent staff turnover, and lower involvement and productivity. On the contrary, psychologically resilient employees quickly adapt to changes, find ways to overcome failures, and are more productive and easier to work with. Psychologically resilient employees communicate effectively, do not avoid risks, and strive for personal mastery; they are characterized by a positive and flexible attitude, constant learning, and self-confidence. The research has confirmed that there is a strong link between psychological resilience of employees and their ability to function effectively. It is important to note that a mentally resilient person is not a person who does not suffer from stress and conflicts, but a person who is able to cope with the requirements placed on him.

In the research aimed at identifying the links between leadership style and resilience of subordinates, usually authentic leadership is described. It is stated that the development of resilience skills is an important component of authentic leadership. As mentioned, a construct of authentic leadership has emerged in the context of leadership efficiency research. It has been noticed that some effective managers cannot be distinguished by any of the existing leadership styles; therefore, the term authentic leadership was introduced.

The theory of authentic leadership emerged a few years ago and it developed from the concept of transformational leadership. Authentic leadership is the behavior of a manager which promotes positive competencies and creates a positive, ethical, and supportive climate (Avolio and Gardner 2005). This behavioral style is characterized by four components: relational transparency, internalized moral perspective, balanced processing, and self-awareness.

*Component of relational transparency.* It is the promotion of open communication; exposure of your true self; open expression of beliefs, values, and views. This behavior builds trust and encourages open information sharing. Open communication leads to strong and close relationships.

*Component of internalized moral perspective.* It is ethical, selfless behavior, based on internal moral standards and values; provision of support and assistance; manager's trust in subordinates; granting of autonomy. In this way, confidence, independence, support, and assistance are promoted and they facilitate adaptation.

*Component of balanced processing.* It is the empowerment of subordinates in decision-making and mutual cooperation, where the manager listens to different opinions and approaches before making decisions. Mutual cooperation leads to a more successful adaptation, flexibility, and faster reactions to stressors.

*Component of self-awareness.* It is the perception of one's strengths and weaknesses, understanding of the influence on other people/subordinates, the perception of one's values, and relationship with subordinates. Authentic leaders are distinguished by openness and consistency in behavior (matching between actions and beliefs, values and views); they behave altruistically in order to direct actions of the group towards a common objective.

Thus, an authentic leader is an individual in a responsible position who is sincere and reliable. The behavior of an authentic leader reflects his values, beliefs, thoughts, and feelings rather than requirements of the environment; this type of leadership is described by such words as true, reliable, trustworthy, and sincere

(Avolio, Gardner, Luthans, and Walumbwa 2005). Self-knowledge and being yourself are important features of authentic leadership. Authentic leadership is the open recognition of one's thoughts, beliefs, emotions, and behavior, reflecting the true self, i.e., telling what a person really thinks and his behavior that corresponds to these thoughts. In this sense, intentions of an authentic leader are open and there is a consistent relation between values and actions. That is why leaders, being role models of company values, have a significant impact on the entire organization.

In addition, the authentic leader is confident, optimistic, honest, and oriented towards the future; he follows moral and ethical standards as well as encourages other employees to become leaders. He enables the conditions for developing trust and people's strengths, encourages employees to follow valuable beliefs, and helps to improve work performance. Such a leader is trustworthy and confident in his subordinates. He supports the values that can make the employee feel the competence and the sense of belonging to a group. In this sense, authentic leadership can be the source of employee development, the basis of health and psychological well-being. It has been established that the psychological resilience of employees at work is associated with the application of authentic leadership principles. Scientists have observed that the implementation of the principles of authentic leadership has such positive effects as better self-awareness and positive self-regulatory behavior of employees. It has been found that employees who perceive a manager as an authentic leader have a stronger sense of well-being and a lower level of stress (Clifton and Harter 2003). In addition to these positive consequences, associated with authentic leadership, such consequences as high self-esteem and psychological well-being, friendly relationships, and improved work performance are also mentioned.

In addition, studies have revealed that employees who perceive their manager as authentic have a stronger commitment to the organization, feeling of happiness, and higher job satisfaction. There is also a positive link among authentic leadership and productivity, the level of customer satisfaction, profit, sense of security of the employees, and the quality of work that is perceived by employees.

Thus, it can be stated that the psychological resilience of employees is the consequence of the introduction of authentic leadership principles in organizations (Cooper, Scandura, and Schriesheim 2005). Authentic behavior of the manager helps to create a healthy work environment, i.e., there are such procedures and systems that help employees to achieve organizational objectives and feel personal job satisfaction. Promotion of psychological resilience is an important task of the authentic leader. The authentic leader encourages psychological resilience of employees by ensuring that they are given support to recover from their failures. Such a leader creates the conditions for thriving during periods of change. He anticipates possible failures and difficulties, creates support plans for employees, and responds to their requests and needs.

In the research on authentic leadership and resilience, faith and the feeling of hope are often emphasized. Faith is the strength of an authentic leader which may affect the psychological resilience of employees who are experiencing changes within the organization. It is claimed that managers can increase the overall resilience of the organization by developing faith in themselves and among their subordinates

(Hajer 2009). There is a significant link between faith, optimism, and psychological resilience.

In conclusion, research results confirm the links between authentic leadership and psychological resilience of employees: authentic leaders develop components of psychological well-being (hope, optimism, and psychological resilience) by encouraging adaptation to changes and unfavorable circumstances. The more developed these constructs are, the more successfully difficulties and requirements are overcome in modern organizations. An authentic leader is reliable and confident in his subordinates.

# 4 Sustainable Leadership

## 4.1 THE CONCEPT OF SUSTAINABLE LEADERSHIP

After analyzing various leadership theories and leadership types, the concept of sustainable leadership may be defined. Sustainable leadership encompasses all major types of leadership and combines them through economic, social, and environmental dimensions embedded in the personal (institutional) dimension, which is significant for the successful assurance of key economic, social and environmental values (Amar and Hentrich 2009; Goffee and Jones 2009). This leadership can be distinguished by its value orientation based on values, mission, and the vision for its implementation. Thus, it includes the most essential principles of leadership based on values, mission, and vision, as well as of moral and spiritual leadership and ensures such important values as environmental protection and preservation of natural resources for future generations. Sustainable leadership is oriented towards people, based on the interaction with the team members and followers, and strives to ensure the continuity of leadership and to train future leaders. It embraces the most important principles of transformational, transcendental, emotional, shared, distributed, and servant leadership as well as ensuring social commitment to the team, community, and society as a whole. Sustainable leadership is oriented towards the objective, which is based on both the mission and values. In this context, it covers the key principles of results-based leadership and enables the organization to achieve its economic, financial, and other development objectives (CISL 2011). Character traits of personality are important for sustainable leadership, as they ensure the combination of all three important components of sustainable leadership: vision and values (environmental protection), interaction with the followers (social dimension), and results (economic dimension). Thus, sustainable leadership encompasses all the key features of charismatic and authentic leaderships as well as the psychodynamic approach of leadership. The dimensions of sustainable leadership and the key components are presented in Table 4.1. Only the coherence and interoperability of these components ensure sustainable leadership.

*Sustainable leadership* is a stable, ongoing process based on the use of non-forced influence on team members, according to the strength and competences of the leader's personality, while trying to direct or coordinate the activities of the

**TABLE 4.1**
**The Components and Dimensions of Sustainable Leadership**

| Components of sustainable leadership | Dimensions of sustainable leadership | Features of sustainable leadership |
|---|---|---|
| POWER OF PERSONALITY | Institutional | Personality character traits, such as initiative, vitality, creativity and preservation, courage, fullness of personality, flexibility, alertness, honesty, self-confidence, balance, independence, autonomy, ambition, perseverance, will, ability to work, need of dominance, aggressiveness, formality, participation, and other important acquired competences are essential for sustainable leadership. |
| INTERACTION WITH FOLLOWERS | Social | Sustainable leadership is oriented towards people and based on the interaction with team members and followers; it strives to ensure continuity of leadership and to train other leaders. |
| VALUES, MISSION, VISION | Environmental | Sustainable leadership is focused on the objective, which is based on the vision, mission, and most importantly, sustainable (environmental, social, economic, etc.) values of a sustainable organization. |
| RESULTS | Economic | Sustainable leadership is objective-oriented and strives for concrete results. |

*Source:* created by authors.

leader's team members as well as to form followers-leaders, in order to achieve an objective of sustainable development of the organization, which is based on the creation of the organizational culture, its values, mission, and vision (Doppelt 2010). Although some authors view leadership as a set of characteristics, attributed to someone who understands that they can successfully use such non-forced influence, the definition of leadership as a process is preferable and more scientifically justified (Newell 2002).

Furthermore, the concept of sustainability or balance implies an organization's ability to adapt to changes in the business environment, to apply its best practices as well as to achieve and maintain competitive advantages, while protecting the environment and natural resources and caring for the company's employees along with the local community (Bartelmus 2013). When creating a sustainable organization or transforming an organization into a sustainable one, some managing practices and managerial competences are required. Therefore, a sustainable organization can be characterized by a high organizational culture, which is also best developed through the principles of sustainable leadership. It is significant to discuss what competences are necessary for a sustainable leader-manager based on the concept of sustainable leadership.

Competence can be defined as a set of knowledge, abilities, skills, and attitudes that is necessary for an individual to work effectively in a specific professional activity (Barth and Busch 2006). It could be:

- professional competence;
- social competence;
- conceptual competence;
- procedural competence.

Professional competence is specific knowledge and skills in the field of work activity, as well as the understanding of processes and technologies, market and competitors, or production and service areas. Social competence is the ability to communicate and work with people. It is based on the features and abilities of an individual to adapt to the social environment. Conceptual competence is systemic thinking, the ability to model situations by employing extensive knowledge and experience, as well as a clear understanding of ongoing processes. Procedural competence is the manager's ability to set priorities, foresee directions of activity which should be followed, steps which should be taken, and methods that need to be adapted. These competences are increasingly becoming strategically important in the business world, which is particularly relevant for managers since competences allow organizations to gain an advantage due to the human factor (Bereiter and Scardamalia 2003).

Scientists have formulated four groups of strategically important competences that outline guidelines for developing competences of an individual in relation to the use of career opportunities (Crofton 2000):

1. Competences of a good manager *(inspiring motivation; initiative; empathy; self-presentation).*
2. Possession of a vision *(strategic management; openness to changes; innovation; decision-making).*
3. Purposefulness *(objective achievement; procedural competency; learning and development; knowledge management).*
4. Cooperation *(teamwork; communication; conflict control).*

In order for business development to be successful, personal intuition or business flair is no longer enough. It is important to be able to find the tools that will allow employees to reveal their potential, encouraging them to take responsibility. The manager has to motivate and support employees, raise awareness of their value, notice their achievements, listen to employees' problems and help to deal with them, as well as to behave constructively when the employee makes a mistake. In this way, the manager creates sustainable relationships across the organization. A fundamental difference exists among agreement, obedience, and devotion, which is the willingness. Only the leader is able to achieve the willingness of others to fulfil his wishes and to be committed to work. Leaders do not need to control and monitor their employees because they follow the leader themselves.

On the basis of the analysis of leadership theories and the concept of sustainable leadership, it can be stated that sustainable leader has to be well aware of the organization's mission and to have a clear vision of the organization that reflects the organization's sustainable values, such as environmental care and preservation of natural resources, concern, and social obligations for employees and the local community.

However, merely having a vision does not ensure the success of the organization and its sustainable development; thus, the purposefulness of activities and management is needed. Only purposeful movements towards the selected objective allow the individual to reach it. Personal features, such as determination, consistency, etc., are especially important for this. When the environment is changing, it is very important to realize the ultimate objective. Moreover, it is important to flexibly react to certain situations. The aspiration to achieve the objective is also related to continuous learning and development since a positive attitude towards learning and development leads to the acquisition of not only new knowledge and skills for reaching the objective but also relevant competences.

The ability of both managers and employees of the entire organization to cooperate, achieve common goals, aid one another, effectively communicate, and work well in a team leads to sustainability in the organization. However, in the collaborative process, many obstacles interfering with the effective exchange of information exist (Keulartz, Korthals, Schermer, and Swierstra 2005). These can be viewed as the more human aspects of communication: interorganizational relations, values, and certain provisions. Communication skills are the ability to establish psychological contact, actively listen, clearly articulate thoughts, and actively communicate. When creating relations, empathy—the ability to feel another person's emotions and needs—is important, as well as an active interest in other people's concerns. Leaders who have social awareness provide emotional support for their employees and, thus, help subordinates to fulfil their job requirements. Emotional support is important to overcome stress and negative emotions as well as to develop courage, enthusiasm, and pride in work that contributes to the development of sustainable relationships within the organization.

In summary, it can be said that sustainable leadership is a process based on the use of non-forced influence on team members, founded on the key competences of a sustainable leader: good management competence, cooperation, purposefulness, vision creation, and teamwork, as well as the development of the leadership competence in a collective. Thus, the concept of sustainable leadership is a holistic concept of leadership and covers the most important dimensions of sustainability and all the key components of the leadership process (power of personality, interaction with followers, values, vision, and mission, as well as results, and the most important types of intelligence: rational, emotional, spiritual, and physical).

## 4.2 PRINCIPLES OF SUSTAINABLE LEADERSHIP

Sustainable leadership was discussed by Andy Hargreaves and Dean Fink (2006) in the book *Sustainable Leadership.* They provided a definition of sustainable leadership for education. "Sustainable educational leadership and improvement preserves and develops deep learning for all that spreads and lasts, in ways that do no harm to and indeed create positive benefit for others around us, now and in the future" (Hargreaves and Fink 2006).

They distinguished seven educational changes and principles of sustainable leadership which can be applied to the consolidation of sustainable leadership in the organization (Hargreaves and Fink 2006):

1. Sustainable leadership matters. The first and key principle of sustainable leadership is to be a leader while learning and taking care of others.
2. Sustainable leadership lasts. It preserves and continually enhances the most valuable aspects of life year after year, moving from one leader to another. The challenges of succession and continuity of leadership are the basis for sustainable leadership.
3. Sustainable leadership spreads. It supports the leadership of others and depends on it. In a complex world, no leader, institution, or nation can control everything without the help of others. Sustainable leadership, first of all, is distributed and shared leadership.
4. Sustainable leadership does no harm to and actively improves the surrounding environment. It does not "steal" from others. It does not try to thrive at the expense of others. It does not damage other organizations but actively seeks ways to share with them its knowledge and resources. Sustainable leadership is not egocentric; it is socially just.
5. Sustainable leadership promotes cohesive diversity. Strong organizations promote diversity and avoid standards that weaken learning, adaptability, and resilience not only to those changes and threats. Sustainable leadership promotes teaching and learning diversity and learns from it. It encourages organizations to move forward by combining completely different components.
6. Sustainable leadership develops and does not deplete material and human resources. It takes care of its leaders by encouraging them to take care of themselves. It renews people's energy resources. Sustainable leadership is wise and smart leadership, which does not waste either its financial or human resources.
7. Sustainable leadership honors and learns from the best of the past to create an even better future. In the chaos of changes, sustainable leadership firmly protects and retains its long-term goals. Many theories of change refer to changes, despite or without remembering the past. Sustainable leadership "refreshes" the memory of organizations and respects the wisdom behind it as a way of learning, preserving, and improving.

In their book *Sustainable Leadership*, published in 2006, Richard Boyatzis and Annie McKee state that sustainable leaders create a powerful recipe for organizational activity by using financial, human, intellectual, environmental, and social capital. It is based on attentiveness, hope, and compassion, which create a harmonious relationship while at the same time guaranteeing a cycle of renewal. In their book, Boyatzis and McKee review the research that has been conducted over decades in various fields of science and provides a practical guide for leaders on how to build and maintain coherence in their relationships, team, and organization (Boyatzis and McKee 2006). According to the authors of the book, in order to overcome the inevitable "power stress," managers need to get rid of the destructive routine and regain their physical, mental, and emotional powers by consciously managing their "cycle of sacrifice and renewal." As proven by R. Boyatzis and A. McKee, the best leaders not only are motivated themselves, but also are able to spread positive attitudes and inspire the surrounding people. True leaders know that they are also guided because leadership is a road of mutual traffic. Any manager must listen to and try to understand others in order to notice the signs that help in his hard work. In the book *Sustainable Leadership*, convincing examples from the world's most prominent organizations are provided, showing how three key elements—mindfulness, hope, and compassion—are essential for renewing and preserving coherence in the collective and the organization (Boyatzis and McKee 2006). These concepts have emerged as the foundation of practical leadership, leading to positive physiological and psychological changes, which in turn enable managers to overcome the negative effects of chronic stress. By examining mindfulness, hope, and compassion, scientists understand mindfulness as a conscious understanding of themselves and others, which, by using the power of emotions, develops intelligence as well as cares for the body and spirituality. Mindfulness is a state in which we are awake, aware, and attuned to ourselves and the world around us. Hence, hope and compassion are feelings, and the experience of them is the basis of leaders for creating harmonious relationships with those around them. Hope raises the spirit and concentrates the energy; because of it you want to take action and use your own strength to achieve the objective. Hope and compassion are contagious and affect the behavior of others. Sustainable leaders live according to their values and sincerely take care of other people. They bring hope to the future and give enthusiasm to the present for themselves and others.

According to the insights of Andy Hargreaves and Dean Fink, and Richard Boyatzis and Annie McKee and the established concept of sustainable leadership, seven universal sustainable leadership principles that apply to both public and private sector organizations, seeking for sustainable development or the transformation into a sustainable organization, can be distinguished (Boyatzis and McKee 2006; Hargreaves and Fink 2006).

1. *Sustainable leadership creates and preserves the long-term sustainable activities of an organization.*

   In any area, the first principle of sustainability is to create what is sustainable and harmonious. To preserve means to foster. Therefore, the activities of a sustainable organization create and preserve what is important and significant, as well as promote the long-term progress and involve the organization's employees into this process intelligently, socially, and emotionally. This is not only about achieving particular results. These are activities that do not focus on achievements but on what is significant in the context of fundamental values of the organization and its mission. The main task of sustainable leaders is to develop the organization's activities in the long-term perspective, instead of taking care of short-term financial results and profits. Sustainable leadership, which strives for the sustainable functioning of the organization, goes beyond the pursuit of the temporary benefits that result from the rapid improvement of performance. Sustainable leadership is first and foremost directed towards ensuring long-term performance, which also greatly benefits society, the organization, and its employees.

2. *Sustainable leadership promotes long-term progress.*

   The takeover of leadership is the latest challenge for leadership. It is the ability to refuse, continue, and plan even when you have to leave. Sustainable achievements are not short-term changes that disappear when there are no heroes left. Sustainable leadership is not intended only for charismatic leaders whose duties are difficult to take over. This leadership is spreading through the exchange of other members of the collective through the influence chains that link leaders' actions with the actions of their predecessors and successors.

   The takeover of leadership almost always has an emotional tone and is accompanied by feelings of expectation, anxiety, loneliness, loss, or relief. The frequent and repetitive takeover of the position only strengthens these emotions. Therefore, the prospective manager should be prepared to take over the position from the first day of the leader's appointment. In addition, the pace and frequency of takeovers must be assessed and established so that employees would not suffer from the cynicism caused by tiredness due to the constant change of leaders. If in the organization several managers change within a few years, employees become very cynical. This principle of "revolving doors" or the "carousel" type of takeover of leadership is increasingly spreading by implementing reforms in risky situations that can be compared with the demographic crisis, where the key leaders of that generation are lost due to their retirement. Sustainable leadership must focus on the takeover of leadership. In order to successfully implement the takeover of leadership, it is needed to (1) prepare the future successors in order to ensure continuity; (2) maintain the successful leaders in the organization for longer if they undertake major transformation to promote change; (3) resist the temptation to look for irreplaceable charismatic heroes; (4) seek that the plans of managers would be included in organizational development plans; (5) reduce the change of leaders.

3. *Sustainable leadership strengthens the management of others and develops new leaders.*

   One of the ways for leaders to create a long-lasting value is to ensure that leadership is developed and shared with other people. Therefore, the change of managers means more than the training of successors. It is also the ability to distribute leadership between the members of an organization or community in order to maintain and further develop ideas after the former manager leaves; in this way, the situation concerning the change of managers is mitigated in the community. Disappointment and the constant change of leaders encourage employees to join the union. The more this union becomes stronger, the more managers are successful in "putting down roots" in this organization since all new managers are appointed with the hope that they will "withstand" this union. However, sustainable leadership is not only the responsibility of individuals. In a very complex world, no leader, no institution nor nation can control everything without the help of others. Sustainable leadership can be viewed as shared difficulties and collective responsibility.

4. *Sustainable leadership is socially just.*

   Sustainable leadership benefits not only a few organizations or a certain group of employees at the expense of others but all organizations and society as a whole. Sustainable leadership is an interconnected process. People who implement it acknowledge and take responsibility for making the organizations influence each other through mutual cooperation. In this respect, sustainability and the takeover of leadership are inseparably linked to social justice and guarantee it. Therefore, sustainable leadership is not only related to the improvement of activities in your own organization. It is the manager's responsibility for the actions in the managed organization and for its employees, as well as for the wider environment, relationships with other organizations, and their influence on each other. This means social justice.

5. *Sustainable leadership encourages development of human and material resources and does not just waste them.*

   Sustainable leadership includes inner reward (satisfaction) and external incentives which help to attract and maintain the most professional and talented leaders. Leaders have time and opportunities to communicate with colleagues, learn and support each other, exchange information and experience, as well as to develop and watch over future leaders who will take over their duties. Sustainable leadership is cost-effective but not too close-fisted. It gives opportunities to carefully use and save available resources by developing talents of all the employees rather than wasting these resources on the selection and rotation of the already "prominent stars." Sustainable leadership is based on a management system that takes care of the managers and teaches them how to take care of themselves. Leaders of organizations are "exhausted" through excessive demands and decreasing resources, and they have neither the physical energy nor the emotional ability to develop professional learning communities. The emotional health of leaders is an asset that needs to be preserved. Leadership that exhausts leaders does not last long. If no one takes care of leaders' personal and professional well-being, the organization will reach only short-term benefits and will threaten the entire future of leadership. Only the most motivated and committed leaders can stay under such unfavorable conditions for a long time. During these reforms, managers can choose early retirement, can be admitted to hospital due to the unbearable pressure, or can leave the position of a leader and move to a lower level, where they are able to deal with the appointed tasks. Thus, sustainable leadership is a leadership that provides support and values leaders.

6. *Sustainable leadership encourages the creation of a diverse environment and develops skills of community members.*

   Individuals who chose sustainable leadership develop and create an environment where continuous improvement can be promoted on a large scale. They allow people to adapt to an increasingly sophisticated environment and thrive in it by learning from each other's experience. Standardization is the enemy of sustainability. Sustainable leadership creates the prerequisites for the recognition and development of a variety of teaching and leadership methods and provides the opportunity to share experience and practice with other participants of the development process. This is not the imposition of standard templates on all individuals.

7. *Sustainable leadership actively impacts the environment where the organization is located.*

   Sustainable leadership will be consolidated in organizations if their leaders commit themselves and give enough attention to community support, as well as are able to pursue their objectives with determination. In addition, they will take action with a strong vision and will rationally distribute their energy in order not to "overstrain," while achieving their objectives. Such leaders will be able to consolidate improvements that have been made so that these changes would remain for a long time, even after these managers leave their positions. Sustainable leaders are aware of the impact their leadership has on surrounding organizations and the local community; thus, they protect themselves and encourage others to protect the environment, natural resources, and ecological diversity. They do not follow standard instructions on how to lead an organization. Many managers try to implement what is important and to inspire other people to work with them as well as to find their successors, who will lead the company when they leave their positions. Sustainable leadership must undoubtedly become a commitment to all leaders. If the changes in sustainable development that are spreading and last for a long time are important for the society, sustainable leadership must become a guiding principle in the systems where leaders have to do their job.

## 4.3 INTELLIGENCE AND SUSTAINABLE LEADERSHIP

The concept of sustainable leadership is a holistic concept of leadership which covers the most important dimensions of sustainability and the components of the leadership process, together with all the key types of intelligence: rational, emotional, spiritual, and physical (Gottfredson 1998).

The human psyche consists of several or even a dozen sub-areas. Aspects and abilities of these sub-areas of human mental functioning are sometimes also referred to as separate types of intelligence: cognitive, social, musical, emotional, kinesthetic (control abilities of your body's system of bones and muscles), practical (daily life skills), verbal (language skills), and non-verbal (spatial construction skills). Thus, more than ten of such intelligences can be counted. All of them form a coherent, indivisible whole, which is divided into separate species only for didactic, educational, or other reasons.

Physical intelligence of physical quotient (PQ) reflects the ability of a human to listen, identify, and respond to his physical body needs, pain, hunger, depression, and frustration, and to be able to relate the body to the mind, stop eating at the right time, rest, as well as maintain the ideal weight, physical capacity, and health of your body (Pfeifer and Bongard 2006).

Because of rational intelligence called intelligence quotient (IQ) or simply intelligence, we know the world, can identify different patterns, and understand the relationship between cause and effect (Banerjee et al. 2009). The flexibility of our thinking and other cognitive processes determines how fast we process information and make decisions. Since the beginning of the twentieth century, human intelligence has been equated with an IQ. Psychologists of that time created a test to find out the individual's IQ. Such tests were started to be used to select the best and smartest ones. They had initially been employed by the US military to select future officers, then universities and employers started using them to find individuals who are ambitious and talented, and have great aspirations. The results of IQ tests show some basic and often hereditary (at least it was thought so) spatial thinking, mathematical and linguistic abilities; however, since it was the only measure of intelligence that could be calculated, it was used to assess the intelligence of a person in general. In the 1960s, IQ tests and their results started to be questioned. First, it was understood that the test can only assess a certain type of intelligence—rational and logical intelligence, which is needed to solve logical tasks or is used when strategic thinking is required. This type of intelligence is developed by Western education systems, and it prevails in Western business. Second, based on research carried out in the 1960s, psychologists found out that the results of the IQ test of different ethnic groups and different genders were different or varied (Burkart, Schubiger, and van Schaik 2016). Psychologists concluded that either the test is poorly formed or distinct levels of intelligence are specific to different ethnic groups, races, and genders (Jaeggi, Buschkuehl, Jonides, and Perrig 2008). While both options might seem arguable, the IQ test was the only measure of intelligence at that time.

But thinking does not work separately from other intelligences. If we want to think, we must get some kind of incentive. Thinking requires constant reinforcement, which is associated with emotional and motivational mental sub-areas. From the point of view of mental historical development, emotional intelligence—human emotional responses and the ability to regulate them—is one of the oldest intelligences. Expression of emotions is a key mechanism for regulating animal relationships. Later, in the course of civilization, higher feelings, civilized emotions, emerged. Due to the evolutionary adaptation in human society, altruistic tendencies have developed. Automatically, a mechanism that stops people from non-altruistic activities (e.g., shame, guilt emotions) had to appear as well. The emotions of interest (curiosity, desire to understand) encourage the search for innovations, i.e., cognition becomes active and in this way is developed. Cognition, combined with emotions, makes people socially adaptive, i.e., capable of communicating with others, creating friendships, love, cooperation, and other relationships. The concept of wisdom was changed only in the middle of the 1990s with the appearance of Daniel Goleman's book about "emotional intelligence" or emotional wisdom. In this book, the concept of emotional intelligence or emotional quotient (EQ) was formulated. D. Goleman relied on studies conducted by neurologists in the most famous American universities. They discovered that emotions are an important factor in human intelligence (Goleman 1995, 1997, 1998, 2000). If our emotions are healthy, mature, and not affected by any brain trauma, we will effectively use our IQ no matter what it would be (Chopra and Kanji 2010). However, if our emotions are distorted and immature or if the emotional center of our brain is damaged, we will not be able to use any of the available IQ wisely or properly. If the processes which help us to feel are disrupted, our thinking effectiveness is greatly reduced. An illustration of this statement, a classic example, is a distracted, incredibly talented, and intelligent professor who is unable to tie his shoelaces. Another example can be individuals with a supposedly lower level of intelligence (IQ) who have been awarded the excellent ability to work with other people. Thus, emotional intelligence reveals how we interact with other people, how we understand them, as well as situations in which we face them. It also shows our ability to perceive and manage our own emotions related to fear, anger, and aggression or regret. According to D. Goleman, if we are unable to control our emotions, they will control us. D. Goleman expanded the EQ concept by arguing that emotional intelligence is our ability to assess and recognize the situation, understand the emotions of other people and ourselves, and behave accordingly. It can be argued that D. Goleman's work has fundamentally changed the concept of intelligence and has been quickly adopted in practical life and workplaces (Goleman, Boyatzis, and McKee 2002). Unlike IQ, which essentially remains unchanged throughout life, EQ can be developed and improved. We can learn how to behave more wisely with other people and how to control our own emotions. EQ has also slightly broadened our understanding of strategic thinking, as it has turned out that people act following not only their mind but also their emotions, while rational strategies that we create often hide many emotions. However, the EQ concept was not the end of human intelligence research.

In the late 1990s, neurological studies revealed that the brain has another, absolutely different "quotient" of intelligence. It is an intelligence which helps us to understand and feel the innermost meaning, the most important values, the real-life objectives, and to perceive how significant this meaning, values, and objectives are for our lives, actions, and reasoning (Zohar and Marshall 2000). It is called spiritual intelligence, or spiritual quotient (SQ). In their book *Spiritual Capital*, which was published in 2004, D. Zohar and I. Marshall described in detail the concept of spiritual wisdom or spiritual intelligence and revealed its key qualities and interactions with other types of human intelligence (Zohar and Marshall 2004).

Spiritual intelligence encourages us to look for the answers to the most important questions: Why was I born? What is the meaning of my life? Why do I dedicate my life to this relationship, to this work, or to this matter? What do I really seek in my life? Spiritual intelligence allows us to see the broader context of events and create a more detailed picture. As the sky, it encircles our life with a sense of meaning and values. Due to spiritual intelligence, we experience something much greater than ourselves, something incomprehensible that gives sense and meaning to our limited life. Psychological surveys have revealed that at some point in their lives as adults, 50–70 percent of people have consciously felt such a touch of infinity: a moment of total beauty, the deepest love, the most truthful truth, or a dignified sense of unity of all beings and things.

In addition, spiritual intelligence helps us to improve (Zohar and Marshall 2000). It not only keeps what we already know or can do but also encourages us to go into the unknown and experience what might be. It invites us to seek and rely on higher incentives. In the evolution of the human species, seeking of meaning led the brain to develop language. With the evolution of human society, meaning and the search for the most important values made us select the best group leaders who inspire us to dream and strive (Reave 2005). The search for even greater values, meaning, and purpose, which is promoted by spiritual intelligence, makes us feel unsatisfied. Finally, spiritual intelligence gives us unlimited insight and understanding of the whole situation, problem, or totality of being. It allows us to clearly perceive or discover the depth and meaning of things. It allows us to change the rules or to create new ones. It allows us to criticize what already exists and to dream of what might be. This intelligence allows us to imagine situations or opportunities that do not exist yet. It is the intelligence of transformation, destroying the old totality and creating a new one. In this way, we can get rid of old business models and old ways of thinking, explore problems and situations in a new, broader context, abandon old incentives, and move towards higher motivation. Thus, spiritual intelligence is the basis for strategic thinking that helps us to look at our strategy from the side and reassess it.

Various leadership theories which were examined highlighted the importance of separate types of intelligence to ensure the implementation of a particular type of leadership. Meanwhile, the concept of sustainable leadership can be seen as a concept of holistic leadership that covers all four key areas describing the essence of human existence at work: body (physical), mind (logical/rational thoughts), heart (emotions, feelings), and spirit. Previous leadership theories focused on the different degrees of physical, mental, and emotional dimensions of human communication within the organization but did not include the concept of holistic leadership.

In Table 4.2, four types of intelligence are presented, which are important for the implementation of sustainable leadership principles.

Thus, the basis of sustainable leadership is physical, rational, emotional, and spiritual intelligence. These intelligences cover the main dimensions of sustainable leadership and enable sustainable interaction and synergistic effect of sustainable leadership as a process of key components (power of personality, orientation towards people, value-orientation, and orientation towards the objective and results) (Fry 2003; Fry, Matherly, Whittington, and Winston 2007).

In the following section, we will discuss the concepts of sustainable leadership and reveal what specific knowledge, skills, and types of intelligence are characteristic of sustainable leadership, as in order to achieve the sustainable development, implement principles of sustainability at the organizational level, and transform them into sustainable organizations, it is necessary to use the principles of sustainable leadership which would rely on economic, social, and environmental values, would implement the mission of the organization, and would guarantee its long-term development.

**TABLE 4.2**

**Types of Intelligence Together with the Dimensions and Components of Sustainable Leadership**

| Sustainable leader | Needs | Types of intelligence | Q | Characteristics | Vocation | Components of sustainable leadership | Dimensions of sustainable leadership |
|---|---|---|---|---|---|---|---|
| Body | To live | Physical intelligence | PQ | Discipline | Need (to see emerging needs) | Personality power | Institutional |
| Mind | To learn | Rational intelligence | IQ | Vision | Talent (the volitional focus of attention) | Results | Economic |
| Heart | To love | Emotional intelligence | EQ | Enthusiasm | Enthusiasm (work satisfaction) | Interaction | Social |
| Spirit | To leave a footprint | Spiritual intelligence | SQ | Conscience | Conscience (to do what is right) | Vision and values | Environmental |

*Source:* created by authors.

## 4.4 CHARACTERISTICS AND COMPETENCES OF A SUSTAINABLE LEADER

When analyzing the personality of a leader and its influence on sustainable leadership, physical, rational, emotional, and spiritual intelligences are significant. The manager has to be educated and have certain intellectual abilities in order for leadership to be useful for everyone (Buford 2001).

Physical intellect allows a sustainable leader to stay healthy, energetic, and diligent. With the help of physical intellect, a person is able to understand and identify the needs of his physical body, pain, hunger, depression, and frustration; in addition, he is able to relate the body to the mind, stop eating at the right time, as well as maintain the ideal weight, physical capacity, and health of his body, which is a necessity for a sustainable leader.

Rational intellect provides the sustainable leader with the following competences and abilities: mind, ability to think logically, education, knowledge of the area of activities, wisdom, ability to understand the essence of an issue, erudition, conceptuality, intuition, originality, linguistic abilities, and the desire to know.

Emotions are also very important in the life and activities of leaders. It is known that no matter how carefully work is planned, or all processes are balanced, it is still impossible to avoid emotions in certain situations. By identifying and understanding the emotions of themselves or others, leaders can increase their competences (Zeidner, Matthews, and Roberts 2004). Emotional intelligence is an ability to control yourself and your relationships with other people. Managers capable of redirecting emotions of subordinates in the right direction receive the most benefit. Emotional intelligence allows an individual to perceive, compare, and understand emotions as well as to control them. Emotional intelligence can be described as the ability to perceive and express emotions, use them to facilitate, understand, and reason thinking, as well as efficiently control them in yourself and in relationships with others. Primarily, this intelligence is an ability to perceive emotions, to understand information which is coded behind those emotions, and to control them properly.

Emotional intelligence consists of four areas (Collins 2001):

- self-awareness;
- self-management;
- social awareness—empathy;
- relationship management.

The first two areas determine how people are capable of understanding and controlling themselves and their emotions. Social awareness and relationship management determine the capability of identifying and controlling the emotions of others, creating relationships, and working in complicated social groups. Every area of emotional intelligence (self-awareness, self-management, social awareness, and relationship management) supplements the basic abilities of sustainable leadership. Therefore, it is very important for sustainable leaders to control their emotions because work performance is significantly influenced by feelings. It is important to identify reasons for the emergence of anger, why it occurred, and to find appropriate solutions.

Self-awareness is an understanding of emotions, strong and weak characteristics, values, and motives. People who have self-awareness are loyal to their principles, realistic, frank with themselves, treat others honestly, and have their own values, objectives, and dreams. An important characteristic of self-awareness is a tendency to self-analyze and reflect. Such people act confidently and that is one of the necessary prerequisites for sustainable leadership. A person with good self-awareness also has a very good intuition that allows him to make the best decisions based on experience (Weinberger 2003). The best case is when decisions are based on both intuition and available information. Intuition allows emotionally experienced managers to use the experience of life, while perfected self-awareness enables this to be properly perceived.

Self-management is an ability to control emotions and impulses. First of all, it is important to understand what those feelings mean; only then people become capable of controlling emotions. With the help of self-management, people do not give in to feelings. Only managers who have strong self-management spread optimism and good mood, are energetic and in such a way create a good micro-climate in the organization. Uncontrolled emotions of a manager have a negative impact on the work of his subordinates because, as is known, a bad mood is contagious (Wong and Law 2002). Of course, the manager is an individual too; therefore, it is possible for him to have negative emotions in personal life; however, it is significant that those personal issues do not affect the work relationships of the manager. Managers capable of perfectly controlling their emotions are able to adapt to changes more easily and also help the organization to adapt.

Social awareness is an ability to feel the emotions of other people, to perceive the situation from a different perspective, and to actively show concern for others. Moreover, it is an ability to plan the sequence of actions, find a way to make decisions, and understand the policy of the organization. It is an ability to understand and fulfil the needs of employees, customers, or clients. Social awareness impacts how a leader will achieve the mission of the organization. By knowing how others feel, it is easier for the manager to understand others and how to behave in particular situations. An ability to hear out others and look at the situation from a different perspective helps the manager to understand the emotions of employees and to create understanding and support for each other in the team. Only by understanding emotions of people and managing interrelationships can a leader be capable of achieving the mission of the organization, make an impact on others, and make sure that they participate in the achievement of the mission or the realization of vision and objectives.

Relationship management is a perfectly planned control of the emotions of other people. It is a combination of self-awareness, self-management, and social awareness. To be able to manage relationships means to be capable of persuading, controlling conflicts, and successfully cooperating. Sustainable leaders have to understand their own social awareness and consciously interact with subordinates. Relationship management is based on sincere behavior and feelings of a manager. Only by following their own vision and values can managers spread positive emotions in the organization, be capable of adapting to the emotional environment of a group, as well as direct their subordinates and show them the right way. Emotional intelligence helps a person to understand certain situations and act accordingly.

Emotionally intelligent leaders, first of all, try to understand themselves, who they are, and perfectly control their emotions; they create sustainable and trust-based relationships around themselves. It is known that emotions are contagious; thus, emotionally intelligent leaders understand that fear and anger can mobilize people only for a short period of time; however, such emotions quickly backfire, as subordinates become distracted, restless, and incapable of working. Emotionally intelligent leaders are determined, caring, and passionate, and inspire employees for a common objective. Such people want to act, work together, and move forward. Emotionally intelligent leaders give courage and hope to others and help to reveal the best qualities of employees. A manager who has mastered skills of cooperation, keeps a high level of cooperation in the group and ensures that members of that group make decisions which lead to positive results. Such managers are capable of maintaining good interrelationships with employees; however, together they create not only a friendly but also a working atmosphere (Thor and Johnson 2011).

Therefore, it can be said that skills of emotionally intelligent managers are the incentive to achieve better results, take initiative, have good skills of cooperation and teamwork, as well as a great ability to manage a team. Emotions are very important in the life and activities of leaders. No matter how well these activities are performed, it is impossible to avoid emotions at work. Only by understanding and assessing emotions of themselves and others can leaders increase their competence (Dulewicz and Higgs 2000).

Another important aspect of sustainable leadership is spiritual intelligence. A manager by using a non-formal method seeks for sustainable leadership which is characterized by a few components. Ideological influence (ability to persuade subordinates to believe in the ideas of their leader) is very important. Inspirational motivation is a collective individualized feeling of community created by the vision of a leader. Intellectual stimulation is an encouragement to subordinates to search for ways of improving their performance. In addition, a spiritual relationship with subordinates is very important. An assumption can be made that spiritual leaders tend to have outstanding qualities that make them unique. The following areas in sustainable leadership that are influenced by spiritual intelligence can be distinguished: personal and interpersonal relationships; problem-solving and objective achievement; motivation, commitments, and responsibility.

*Personal and interpersonal relationships.* Empathy is a necessary condition for spiritual intelligence. It allows leaders to comprehend colleagues and to be more understanding, which is one of the characteristics of spiritual intelligence. Spiritual intelligence improves the ability of a leader to understand others better. A leader with high spiritual intelligence can be distinguished by these characteristics: he does not condemn in front of others; he analyzes factors related to problems of others; with appropriate reasoning, he finds a solution for a situation or event that has led to the crisis. A spiritual leader is characterized by the systematic nature of the carried-out activities. Therefore, it can be said that leaders who have high spiritual intelligence are distinguished by the empathy that, through cooperation, helps to achieve better objectives of the organization and to implement its mission based on values.

*Problem-solving and objective achievement.* Spiritual intelligence helps to find the deepest internal resources of an individual, such as adaptation, concern, and tolerance. Intelligence helps individuals realize their actions and their lives in a more meaningful context. Therefore, one can say that spiritual intelligence helps to solve occurring problems and achieve objectives through three main factors, such as adaptation, concern, and tolerance (Ruderman, Hannum, Leslie, and Steed 2001).

*Motivation, commitments, and responsibility.* Spiritual intelligence develops potential opportunities and contributes to the well-being of employees. An employee has a unique set of skills, abilities, and competences that are the most valuable aspects of human resources. Spiritual intelligence encourages feelings, self-esteem, competences, feeling of completeness, job satisfaction, and inner peace of employees. It provides motivation for work. A happy and satisfied employee is motivated to do his job properly. If employees are happy, they perform well. Therefore, the satisfaction factor of employees is the main aspect of motivation. Based on this, it should be analyzed what provides a feeling of completeness for every individual. In this way, by making a connection between an employee and a leader, the main objectives of the organization are achieved.

Thus, sustainable leadership, based on spiritual intelligence, can be distinguished by certain characteristics: efficiency of management and education; interest in people, their experiences, events, and moments that shape their growth by teaching them and in such a way improve their management skills. Spiritual intelligence contributes to the efficiency of management and organizational performance. Spiritual leadership is distinguished by the aspect of having an interest in people, as management is carried out by using guidance, which creates an opportunity for both sides to improve. Most people find the meaning of being as well as success in work and life through spiritual intelligence. Work brings development and changes in personal, psychological, social, and spiritual life. Growth and improvement are inseparable from the process of change and evolution. Spiritual growth creates a vision and a mission of life. It is a spiritual inspiration of the highest level. Therefore, one can assume that, by developing spiritually, people also improve their emotional intelligence. Based on spiritual intelligence, harmony at work can be discovered as well as the balance between work and home environment.

A sustainable leader with high spiritual intelligence can create procedures and systems that empower his employees and at the same time are team-orientated, easily adapted, consistent, and balanced. A leader also should have clear objectives, mission, and vision of the organization that are shared with his subordinates. Such an environment promotes organizational education as well as reduces the number of complaints, conflicts of roles, and possibilities of friction. Spiritual leadership acts as a mediator that strengthens spiritual intelligence on the level of employees. Spiritually intelligent employees participate in the creation of organizational spirituality together with their leader. This has a positive influence on their motivation because they encourage themselves by improving spiritually. By being strongly self-motivated, employees contribute to the improvement of organizational activities, thereby increasing productivity, while the wage becomes a less decisive factor. Such increase in productivity strengthens the motivation of employees and creates a strong culture of the organization. A flexible and sustainable culture of the organization focused on internal and external forces is a source of high-level productivity of employees.

Therefore, in order to successfully lead others, a sustainable leader has to have particular competences. Various authors distinguish different key competences of leadership. Those competences can be divided into six groups: (1) achievements and activities (orientation to achievements; care of order and quality; initiative; aspiration for information); (2) help and services to others (interpersonal understanding; orientation to clients); (3) influence and impact include a good understanding of the organization and the establishment of relationships; (4) management and improvement/development of others, provision of instructions; teamwork and cooperation, as well as team leadership; (5) cognition, analytical and conceptual thinking, high professionalism; (6) personal efficiency, self-control, and high self-confidence; flexibility and loyalty to the organization.

Leadership can be divided into several levels: meta-leadership—provides the direction and vision as well as inspires followers; macro-leadership—searches for ways to implement the vision and create an organizational culture; micro-leadership—focuses on choosing a particular style of leadership. Therefore, the vision is included in two levels of leadership and is often distinguished as the key factor of leadership. In addition, leadership can be grouped according to the performed tasks that are covered and integrated into to the whole by leadership. The performance of a leader can be divided into nine different categories: leader; manager of communications; observer; distributor of information; representative; entrepreneur; solver of crisis; distributor of resources; and negotiator.

According to the majority of researchers, leadership has two functional requirements: the focus on the organizational objectives and the determination to maintain a high-level of philosophy and satisfaction of the group members. Therefore, one can say that the level of satisfaction of members of an organization depends on the potential of the company's leaders to provide it to them (Edwards, Turnbull, Stephens, and Johnston 2008).

Eight important features of a sustainable leader:

1. sincerity;
2. determination;
3. focus;
4. personal involvement;
5. cooperation with people;
6. communication;
7. desire to achieve objectives;
8. improvement.

*Sincerity.* In a healthy company, there is no distinction between things that are said and things that are sincerely believed in. Values come from the leader and become a part of the daily activities of the organization. It is very important for the leader not to hide his personal beliefs. Leaders of successful organizations sincerely express their beliefs and wishes. They do not avoid showing that they believe in the declared values, on which the mission of the organization is based. In fact, it is done quite emotionally. Therefore, it is essential for a sustainable leader to show an example. Being honest in communication is not enough; it has to be shown through actions. Every decision and every action has to be in line with the philosophy of a sustainable leader and be the expression of his values in practice. All actions of a leader affect his employees. A leader is an example of a culture that is being developed within the organization. It is necessary for a leader to prove his words through actions. Of course, there are no leaders that live up to 100 percent of their ideas; however, if leaders cannot prove even a quarter of their ideas through actions, they are not sustainable leaders. When their speeches become only insincere rhetoric, it is impossible to create a strong culture of the organization, to influence team members, and to involve them in the organization's mission. Spiritual intelligence is a huge asset for a leader in order to build sincere relationships within the organization.

*Determination.* The key talent of a good manager is the ability to make decisions. The ability to find a solution even when there is a lack of information (because no one ever controls the whole information) is the most important feature of the leader and an essential feature of a well-functioning team. It is not recommended to reason for too long. A person can never control all facts and data that would allow neutralizing all risks. Moreover, the analysis of any situation depends on the assumptions of someone. Two people thinking about the same facts can make totally opposite conclusions. Why? Because of the different assumptions they make based on the facts they have. However, a decision has to be made; it is unreasonable to spend an eternity analyzing the facts. The objective of a sustainable leader is to make a decision, instead of postponing it because of endless considerations.

It is essential for the leader to use his intuition. How can a leader make a decision if there is not enough information? Partially, by using intuition. Even though, to some people, intuition might seem unscientific and irrational, individuals who have made the most successful decisions in their lives followed a combination of rational analysis and personal intuition. Here are some methods on how to use intuition successfully:

- Look at the essence of a problem. Do not let a personal opinion or determination be damaged by a flow of information, continuous considerations, and opinions of other people.
- Dispose of what is unnecessary (the infinite pros and cons list) and focus on the core issue. When solving a particular problem ask yourself, "What is its essence? If I do not go into details, what is the most important thing here?" Do not waste time thinking about all peculiarities and complexities of the problem. Focus on the essence.
- Sometimes it is useful to distinguish the main issue and simply listen to what your intuition says: yes or no.

It is important to keep in mind that fear affects intuition. Fear gives birth to self-deception. Sometimes a decision made out of fear might seem as a decision suggested by intuition. If the leader is dominated by fear, he might not choose a decision that deep inside seems the right one. A wrong decision might be confused with an intuitive one because of the feeling of relief that comes after the fear disappears. In order to use his intuition efficiently, the leader has to decide to do what seems right, despite the possible risks.

No matter how smart the leader is, it is impossible to always be right. Sometimes a wrong decision will be made—such is life. If leaders wait until they are absolutely sure before they choose, they might lose their determination. It is important to make a decision and move forward if there is a necessity for it. When a problem occurs, the leader has to take the offense position instead of waiting to be cornered. If the leader does not make the best decision, it is not the end of the world: the situation may recur, and the leader will get another chance. The majority of people are afraid to come out on the wrong side, thus the necessity to choose frightens them. People are afraid of becoming a laughing stock or to feel guilty about something; they are scared to receive criticism or be mocked. In other words, the psychological consequences of the mistake are worse than the practical ones. People have to understand that everyone makes mistakes and it is important to learn from them. Mistakes are an invaluable source of power for the leader.

Leaders of successful companies often delegate the decision-making right to their team. The decision-making right is delegated as low as possible according to the organizational hierarchy, which encourages employees at all levels to think quickly, use their intelligence, solve problems in a creative way, and take responsibility (Ferdig 2007). The following recommendations are useful when making a group decision:

- Give people the opportunity to learn how to make decisions. Clearly establish what decisions employees can make independently and explain that they have to take responsibility for their decisions.
- Have your own opinion but be ready to hear out and accept other ideas. Clearly explain who will make the final decision: the group or you.
- Encourage a variety of opinions and a discussion during the consideration of a problem.

No matter what management style is implemented in the organization, everything must be done openly. People will understand that they are being manipulated if a leader informs them that there will be a consensus even though a decision is already made and only the approval of the team is expected. This will motivate employees to assess everything that is happening cynically and will weaken their interest in the company's affairs. If a leader in a particular situation has to be autocratic, he should not hide it. It is necessary to take responsibility and share success. In addition, it is important to be ready to take responsibility for bad decisions and share the success of the positive ones. A leader will lose the trust and support of his team as well as respect of employees if success is not shared and blame is put on others. Some managers see wrong decisions as great ideas that have been destroyed by its implementers. It might be true; however, a true sustainable leader will take responsibility anyway. In case of success, it is necessary to let subordinates feel like winners too. A successful sustainable leader does not feel the need to be at the center of glory: his personal contribution is quite clear. Rational intelligence of a sustainable leader includes determination and risk tolerance when making a decision.

*Focus.* In order to achieve objectives and implement the mission of the organization, a sustainable leader must decrease the number of priority tasks to a minimum and focus on the most important objective. A leader cannot cover and do everything. The organization also cannot do everything. Therefore, it is necessary to choose the most important target and focus on it. It is useful to make a short list of priority tasks and maintain it in that way. If the leader has more than three priorities, it probably means that there are no priorities at all. In addition, it is important to plan time instead of work. There is an essential difference between organizing time and organizing work: work, unlike time, has no end. In order to be productive, a leader must regulate his time rather than the amount of work. By wisely planning time, a lot of spare moments are found. One way to force yourself to concentrate is to work less. John Willard Marriott, the founder of Marriott Corporation, who made a business of a single restaurant into a gigantic corporation, said that people should have fewer working hours since we waste half of that time. One of the reasons why the majority of people have a hard time trying to focus on essential things is because they cannot decide what should be crossed out from a long list of priorities. A sustainable leader has to cross things from his priority list because otherwise the vision and mission of the organization will not be realized. Rational intelligence is important in order for a sustainable leader to focus.

*Personal involvement.* Founders of big companies always show interest in their organizations' affairs. They do not allow themselves to be distanced from what is happening. Therefore, it is necessary to build interrelationships. One of the priority areas of successful companies is the creation and maintenance of long-term, constructive relationships with customers, suppliers, investors, employees, and the outside world in general. It has nothing to do with the artificial attention that is declared by the majority of companies. It is an entirely different thing. In a successful company, relationships between employees and the organization go beyond the standard logic: "I am being paid for my work." A relationship with the company is maintained even after the end of the contract. An example of that would be people who refer to their previous work by saying "we." Such relationships are formed in high-culture organizations, where leaders devote a lot of their personal time to establish these relationships. Personal contact is needed; if a leader only provides written orders, relationships with employees will not be good. First of all, a sustainable leader should leave his office and be seen getting acquainted with as many employees as possible. A sustainable leader should be approachable. If the relationship between a leader and members of his team are too formal, they will not provide the necessary benefits and will not deliver the expected results in the organization. Sustainable leaders need to monitor their speech tone (they should not use foul language) and not abuse their status (luxury, privileges). Sustainable leaders must know what is happening in the organization. They cannot use a common misconception that, in a developing organization, leaders should distance themselves from all internal affairs. Even though the delegation of duties is necessary, a sustainable leader still has to know about the most relevant projects. The best way to do that is to see everything with your own eyes and to hear with your own ears. A sustainable leader must find out what kind of problems or achievements currently exist and how people feel in the organization. Therefore, not micromanagement but personal involvement is significant. A micromanaging manager does not trust subordinates and tries to control every decision. Such leaders are sure that no one except them can make the right decision. Micromanagement limits the personal improvement of employees. It is possible to participate in everyday activities without suppressing subordinates. Personal involvement has an opposite effect from micromanagement: a leader inspires everyone instead of demoralizing them and subordinates can achieve more than they expect from themselves. Emotional intelligence of a sustainable leader determines his personal involvement.

*Cooperation with people: rigor and subtlety.* Sustainable leaders who create sustainable organizations are capable of finding a balance between rigor and subtlety. They require the team members to achieve extremely high results (rigor) while trying to help them to be proud of themselves and self-reliant (subtlety). It is essential for a sustainable leader *to provide feedback.* It is the least applied element of management. Unfortunately, the nature of a human dictates that the best results are achieved when people see themselves positively. Numerous psychological experiments prove that the objective results of human performance deteriorate or improve depending on how others respond to their achievements. Positive feedback improves the quality of work, while negative feedback worsens it. Unfortunately, managers of organizations rarely express their assessments in regard to the performance of employees. The lack of feedback usually brings such misconception as: "the organization does not think that my work is important." When members of the collective feel insignificant, they stop trying. Leaders cannot act as bosses or critics when their reaction is being anticipated. They need to be teachers and advisers. The process of criticism must encourage employees to improve instead of demoralizing them. It is necessary to encourage employees to follow a successful example of their leader as well as stories of other successful leaders. A good teacher just like a good manager has an attitude that every person can achieve great results and strive for success. Since demanding a good performance gives the view that people are lazy and do not try, thus they have to be forced to perform well, a sustainable leader does not require his employees to work well. A sustainable leader provides subordinates with an opportunity to prove themselves. It is necessary to offer people the opportunity to grow professionally and show confidence in them, together with the belief that they will succeed. Emotional intelligence of a sustainable leader helps to ensure the success of cooperation with people as well as feedback.

*Communication.* Emphasizing the importance of communication should not sound banal. Unfortunately, managers of many organizations are not capable of successful communication and some even do not attempt it. Organizations thrive with the help of communication. Live examples can be used in order to explain to others or find out the objectives and the mission of the organization. Analogies, metaphors, comparisons, and graphical images are powerful tools of communication, which help to express the organizational spirit, values, and culture. It is not recommended to explain everything in a direct and straightforward way. Moreover, it is not important whether analogies are correct in respect of logic. The essence of communication is not logic but the effectiveness of impact. It is beneficial to add some personality in the communication. For example, good books provide a feeling that the author is personally talking to his reader. It is essential to try and create such an effect. There are two main methods of achieving it: opening up and speaking straightforwardly.

People do not like being misled. Moreover, people hate when others consider them idiots. Managers who are incapable of being honest and of opening up lose respect very quickly. In addition, it is important to motivate employees to communicate among themselves. Communication has to take place at all levels and directions. Sustainable leaders must realize that, in their style of communication, they show an example.

- When submitting questions, it is necessary to give time for the employee to come up with the answers.
- It is necessary to regularly hold meetings and prepare at least one important announcement relevant to everyone.
- It is necessary to hear out people that do not agree with the group.
- Sometimes things can be done not according to the plan. Improvised informal meetings are one of the best methods of communication.
- Reduction of formalities is necessary for people to feel comfortable.
- Deception should not be encouraged, and it is important to avoid getting in the middle of them.
- Members of the organization must be encouraged to openly express their feeling and thoughts. If feelings are suppressed, successful communication is impossible.
- Dominance of one or two employees during discussions must be avoided. It is necessary to include the silent team members by asking for their opinion.
- It is also essential to thank those who have touched on unpleasant but important topics for the company.

*Desire to achieve objectives.* Sustainable leaders should never be satisfied with what has already been achieved. It is necessary to work persistently but to not overwork. Working persistently and being a workaholic are two different things. Some work persistently to get something done, while others have an unhealthy and manic need to work. The latter type of work is destructive. Workaholics simply become overworked. Only 40–50 working hours per week can be effective. A person can work 90 hours per week, but inefficiently. More does not always mean better. The desire to achieve objectives is related to the rational intelligence of a sustainable leader.

*Improvement.* Learning and improving your skills is a never-ending process. Continuous learning and improvement are some of the most important characteristics of a sustainable leader. Both professional knowledge and managerial skills should be constantly improved. The erudition of a leader is very important. It should be higher than that of the average graduate of a higher education institution. A sustainable leader needs a broad knowledge of his professional field as well as of culture, art, philosophy, politics, geography, etc.

A sustainable leader must stay energetic. If the leader becomes exhausted, the same thing happens to the organization. Therefore, sustainable leaders must take care of themselves physically, emotionally, and spiritually. Physical, emotional, and spiritual intelligences are significant to a leader. The integration of these intelligences ensures high psychological intelligence, which is necessary for psychological resilience both in work and in life. Therefore, it is important for a leader to get enough sleep and rest. Moreover, it is necessary to have various hobbies. Reading, going to the theatre or concerts, watching films, and having conversations with interesting people develop spiritual, emotional, and rational intelligences. Sustainable leaders must be open to new ideas and challenge themselves with new tasks. Everything needs to be done to ensure that a sustainable leader remains an energetic, improving, lively, and truly alive person. It is very important to enjoy your activities. If people are forced to do things they do not like, eventually they lose energy and suffer "burn-out." One of the best ways to maintain energy is to constantly change. Always try new things or do something differently. Of course, it is more convenient to leave everything as it is. Nevertheless, changes not only require an effort but also provide more energy. Sustainable leaders thrive for changes and initiate them, as they continuously model the organizational vision of the future and implement mission-based management in order to achieve the objectives of the organization together with his followers.

## 4.5 SUSTAINABLE LEADERSHIP MODEL AND ITS IMPLEMENTATION

A sustainable leader is the moderator of cultural changes in the company. Cultural change does not occur on its own and the organization does not transform into a sustainable one automatically. A learning process is necessary, during which leaders and subordinates acquire new knowledge, attitudes, and behavior models until each of them takes responsibility for the organizational mission of a certain level. According to the mission-based management type, created by P. Cardona and C. Rey, the following implementation model of sustainable leadership in organizations is proposed: commitment, cooperation, and change (Cardona and Rey 2008). All three parts of the model are closely interrelated and encompass the most important components of sustainable leadership process (the interaction, the mission and the value-based vision, the results, and the power of the personality).

In these three elements of sustainable leadership implementation (in the cycles of commitment, cooperation, and change), a significant role is performed by the organizational mission. The mission is based on the environmental, social, and other values of the organization. It can be described as the formation of organizational culture by seeking to realize the following vision—the creation of a sustainable organization.

First, the cycle of commitment in the implementation model of sustainable leadership will be discussed.

### 4.5.1 Cycle of Commitment

The initial starting point for sustainable leadership is the means which would stimulate the commitment of the organization's employees. One of the main effects sustainable leadership brings to the organization is the improvement of the relationship between the employer and the employee, the increase of the mutual understanding, and the support to each other (Cameron, Quinn, DeGraff, and Thakor 2006). The principal relations between the employer and the employee, including the external ones (working for money) or the internal ones (satisfaction), can be supplemented with the transcendental relations of commitment (the employees feel that they participate in the mission, which needs to be fulfilled, and, guided by the sustainable leader, they make a commitment).

If the sustainable leader complies with the model of the commitment cycle, presented in Figure 4.1, next to their interest in the job, its pay, and the privileges, the followers begin to feel the commitment to the mission.

The first phase of the commitment involves personal commitment. The first factor which influences the emergence of the ownership culture in an organization is the personal commitment of the leader, despite his or her position in the organizational hierarchy. Contrary to the act of empowering, which differs depending on the person's rank, e.g., the chief leader, the manager, or the production worker, in its essence, the commitment to the mission can be of the same level. When the mission is distributed into common missions, the managers of the highest and of the moderate levels must evaluate their personal commitment to the company's mission and values. Mission distribution is not considered successful until the sustainable leader fully accepts and takes the responsibility of this commitment through a common mission. Therefore, it is not enough for the managers of the highest and of the moderate levels to acknowledge their common missions as "theoretical" necessity. They must commit to their assigned mission by participating personally. In such a way, the common mission becomes the

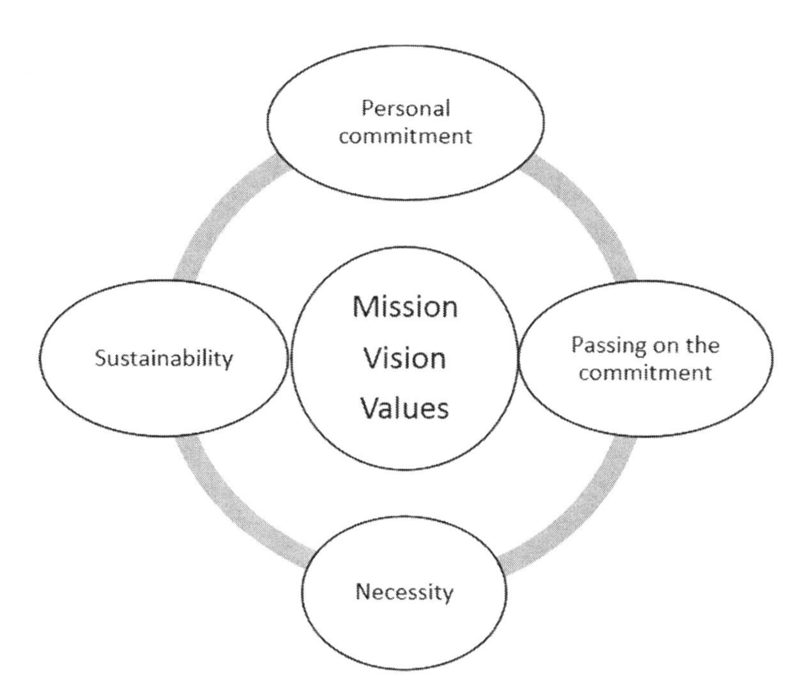


**FIGURE 4.1** Phases of the commitment cycle. (Source: created by authors.)

matter of personal commitment. Starting from the top of the organization, the personal commitment of the leader is a crucial condition for the next phase: to pass on the commitment to the subordinates (Cameron et al. 2006).

The second phase of the cycle is the passing on of the commitment. Generally, employees of the organization consider participation in the activity, related to common mission and values, very attractive. Therefore, the leaders often notice that their subordinates are open to what they are being told and are ready to accept the company's mission and values. However, to strengthen this initial interest and to turn it into a true commitment, sustainable leaders have to demonstrate their own commitment. At first, the subordinates may be skeptical; however, eventually, they will accept the mission and will commit to it, if the sustainable leader displays genuine commitment. To inspire his or her subordinates to make a commitment, the sustainable leader has to pass on his or her personal commitment in such a manner that the subordinates would notice and feel it. To achieve that, the leader has to interact with the subordinates by passing on the information on the mission, the vision, and the values, and by taking advantage of every chance to associate everything they perform with why they do it. In contrast, if the communication revolves around the means to achieve certain objectives, the subordinates will interpret them as the final goals and not as the measures to advance towards the ultimate objective, which is the organizational mission. The leader who concentrates on the goals and not on the means to pass on the commitment does not allow the people to feel the purpose of their work and the connection with the organizational mission. Thus, he or she is not a sustainable leader (Dhang 2012). The sustainable leader can pass on the commitment to the mission only if he or she reveals the essence, which is not limited to the seeking of objectives. With time, the commitment becomes the company's unwritten rule and the necessary condition to integrate into and belong to the organization. The next phase of the process is to link the commitment to the sense of necessity.

The third phase of the commitment cycle is the creation of a sense of necessity. To convert the commitment into excellent work, the sustainable leader has to create a sense of necessity, which would direct the subordinates towards particular, difficult challenges. The sustainable leader possesses various means to create the sense of necessity, including the objectives related to the mission and organizational values, the competences connected with the mission, and the evaluation concentrated on the mission (Dhang 2012). Nevertheless, in order to create a certain sense of necessity, the leader always has to demand the mission-oriented mastery. It greatly differs from the mastery, which is based on profit increase or power demonstration. To demand the mission-oriented mastery, the leader must create the necessity to achieve a certain mission and to develop particular values, since the sustainable leader and the subordinates have committed to it. When the sustainable leader demands the mission-oriented mastery from the colleagues and creates a certain sense of necessity, the combination of the management tools and the leadership is enhanced, and it ensures excellent work for the sake of the organizational mission. However, it does not mean that the leader must follow a set course so that no challenges, arising over time, would force to change the direction or would discourage him or her. This idea leads to the fourth phase of the commitment cycle—sustainable activity.

The fourth phase of the commitment cycle is sustainable activity. As far as the mission and the values are concerned, sustainability signifies two, complementary things. First, it includes the sustainability of various mission components. For instance, if the organizational mission consists of two main commitments to the customers, shareholders, and employees, the sustainable leader must treat all concerned parties with commitment to the same extent, so that no lack of attention is felt to any group for the benefit of another (although in particular moments of time, the priorities will naturally be different). The same holds true for the values. By focusing on one particular value and ignoring the others, it is possible to develop a dysfunctional organizational culture which does not correspond with the organizational values. Second, sustainability must be related to the sustainable leader's perseverance over time. The three discussed behavior models (personal commitment, the passing on of commitment, and the creation of the sense of necessity) are not random or temporary efforts of the leader. The sustainable leader has to implement them resolutely and continuously. If the leader does not act persistently and firmly, it is highly possible that the subordinates will follow the example and all initial efforts to ensure the commitment and to develop the sense of necessity will be wasted. Perseverance is not only the preservation of the commitment. It also includes the renewal and the strengthening of personal and constant commitment, which leads back to the beginning of the commitment cycle. The successful realization of the commitment cycle mostly depends on the sustainable leader's spiritual intelligence.

In the following paragraphs, the cooperation cycle in the implementation model of sustainable leadership will be explained.

### 4.5.2 Cooperation Cycle

Nowadays, cooperation between the different fields, departments, and individual workers of a company is very significant. Businessmen constantly talk about teamwork, cooperation, and internal customers, or compare the organization with an orchestra or sports team. Hence, it may be stressed that, to achieve success, an organization has to execute some sort of cooperation. Nonetheless, after countless decades of attempts to implement cooperation with learning courses, encouragements, evaluation systems, the best employee of the month awards, and similar means, the results are unsatisfactory. The roots of the problem are rather obvious or easy to detect. The lack of cooperation between people is a direct consequence of the absence of motives to cooperate. Generally, it is not related to the shortage of resources, time, or knowledge. People are simply not interested in cooperation. To work together, people need to have a reason or a motive. However, the motives to cooperate are only sustainable if they are transcendental. Cooperation always means to go beyond one's interests, to overcome oneself.

With common missions and mutual dependency, the sustainable leader can create a solid basis for cooperation motives, which encourages attitudes, required for efficient teamwork. Nonetheless, for the system to actually develop cooperation attitudes, the sustainable leader has to ensure that all managers participate in the teamwork process, which is depicted in Figure 4.2 (Dressler 2004).

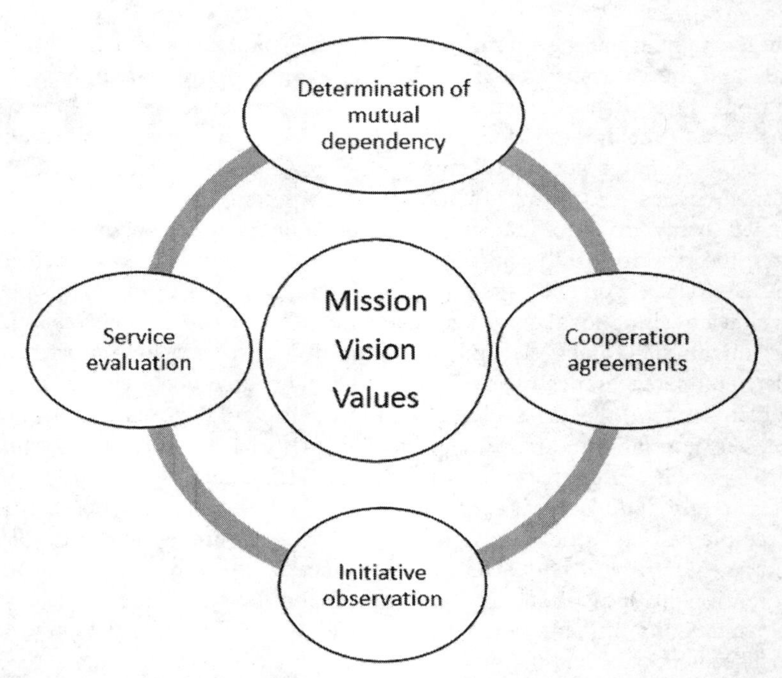


**FIGURE 4.2** Phases of the cooperation cycle. (Source: created by authors.)

The first phase of the cooperation cycle is the determination of mutual dependency. The sustainable leader must know who his or her internal customers are and what their needs are in the context of the company's mission. This aspect differs greatly from the satisfaction of personal needs or working in order to please everyone. The determination of needs in the structure of the organizational mission means that it must be established what a person or department must do in order to more efficiently contribute to the fulfilment of the company's mission. Therefore, in determining the need, the sustainable leader has to understand how that need is connected to the mission of the entire organization. The need for horizontal cooperation can be described as cooperation needs between two or more areas of the company. Horizontal cooperation is based on the "chain of mission's values" and is formed according to the mutual dependencies. This chain reveals how different areas cooperate in order to fulfil the organizational mission. The needs of vertical cooperation are the needs to cooperate between the managers and their subordinates. Vertical cooperation consists of the mutual support between the managers and their subordinates, related to the issues of resources, information sharing, learning, instructing, and other matters required in order to fulfil the common mission of the subordinates. The needs of group cooperation include the needs of cooperation between the employees of the same department. The sustainable leader is responsible for the promotion of the cooperation between his or her team members. Hence, the leader has to encourage team spirit, to avoid inner competitiveness, to stimulate teamwork, to solve conflicts, etc.

The second phase of the cooperation cycle is the creation of cooperation agreements. Having determined the cooperation needs, the sustainable leader has to find a way to successfully fulfil them. Thus, the customer and the provider have to achieve an agreement in the negotiations over their expectations. The sustainable leader has to be open to such negotiations, has to willingly listen to all the parties, and has to determine the priorities of the process related to the organizational mission. In the case of sustainable leadership, this negotiation process must be based on the company's mission and values (cooperation motives must arise from there). Genuine obligations to support another person's common mission should originate from the negotiation expectations of the mission-committed, sustainable leader. These agreements must be explained and stated, preferably in the official organizational documents. In some cases, the commitment may be precisely defined with certain indicators—for instance, a minimal amount of the inventory required to ensure a particular service level in the selling department. However, under other circumstances—for example, in regard to information sharing between departments—a more open commitment may be more suitable. Nevertheless, the essence of such agreements is not control or guilt blaming, but the desire to seek efficiency and effectiveness between the internal activities of the company. To make such agreements truly beneficial, the third phase of the cooperation cycle is needed.

The third phase of the cooperation cycle is initiative observation. To develop a true cooperation culture, it is not enough to only formally determine the commitment between different areas. In his or her team, the sustainable leader must encourage a genuine, mutual-learning culture, based on the initiative observation. Such observation should contribute to the identification of new opportunities in cooperating with various interconnected people. In time, mutual dependencies may change naturally or depending on the internal reorganization process or alterations in the environment. It means that it is not enough to simply fulfil certain collective agreements. Expectations and obligations must be adjusted so that they correspond to the present circumstances and any unsuitable changes need to be corrected. To achieve it, communication channels and trust attitude are needed in ensuring the feedback between the parties. In many cases, organizations have a "non-transparent" and inefficient management, which makes it difficult to prepare feedback. To eliminate such opacity is one of the sustainable leader's main tasks. Representatives of mutually dependent fields should also frequently attend the meetings of the organization's departments. In such a way, each field obtains regular feedback on its internal services level and propositions on how to make the services even more efficient. Thus, the constant sharing of a certain area's mission and the explaining of its needs become second nature to the company and part of the organizational culture.

The fourth phase of the cooperation cycle is service evaluation. The sustainable leader should evaluate each employee according to his or her direct or indirect contribution in fulfilling the organizational mission. Therefore, managers and their subordinates are judged not only depending on their input to their own field, but also according to their input into other areas. Moreover, the assessment must be done by the people who are best qualified in each situation. For instance, internal services must be evaluated by internal customers. The purpose of the assessment is to identify mistakes not in order to find the guilty and to punish them, but in order to improve the services. The analysis allows the organization to move forward and the judgement belongs in the past. The most important thing is to learn to cooperate more efficiently by maintaining the organizational mission as the priority. Then, new needs, which have not been considered or which have not been granted the required priority, may be discovered. Hence, the cooperation cycle returns to its beginning. The successful fulfilment of the cooperation cycle mostly depends on the sustainable leader's emotional intelligence.

In the following section, the change cycle in the implementation model of sustainable leadership will be discussed.

### 4.5.3 Change Cycle

The competencies and talent which the sustainable leader needs in order to skillfully fulfil the organizational mission are constantly evolving, since the needs of the customers, employees, shareholders, and other interested parties are changing. For instance, nowadays the expectations of a car buyer differ greatly in comparison to the car buyer 20 years ago and they will differ even more 20 years from now. The same can be said about the needs and the expectations of employees. The organizational mission can stay unaltered for decades; however, the means to achieve it are constantly changing. On the other hand, one needs to be cautious not to get caught in the opposite trap—change for the sake of change. Sustainable leadership advocates changes on all levels, if they are required by the mission. The process begins with personal change and goes over the change phases, depicted in Figure 4.3.

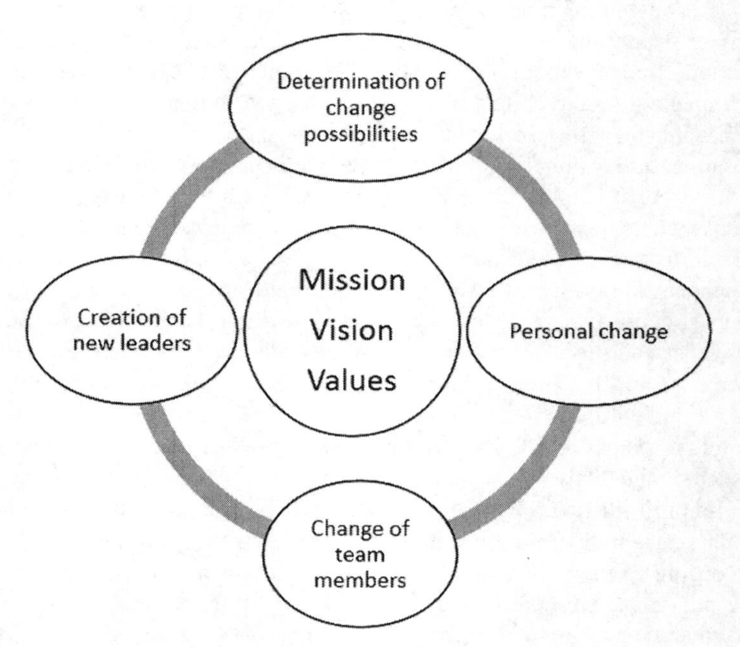


**FIGURE 4.3** Phases of the change cycle. (Source: created by authors.)

The first phase of the change cycle is the determination of change possibilities. When the sustainable leader is personally committed to the organizational mission by feeling the necessity and sustainability, he or she is guided by the genuine ambition which exceeds the momentary interests. It is called the search for mastery. Mastery is the most significant organizational value. By implementing it, the members of the organization always aim to fulfil the mission in a better and more effective way. The commitment to the mission requires mastery and mastery leads towards the change by constantly making effort to develop and improve the organization. In the pursuit of mastery, the change is not the final goal, but the means (it is the most important thing to remember when attempting to perform any kind of change). The search is continuous. It requires a consistent balance between exploitation and exploration. The mastery avoids trivial change and unnecessary strictness. Both of them hinder the fulfilment of the mission. Therefore, in order to achieve mastery, the guide towards the change should be the deep sense of the mission, by attempting to more effectively fulfil the organizational commitments to the customers, employees, shareholders, and other major interested parties.

It is true that the mission provides fundamental stability to the organization's identity and development; however, this kind of stability should not be mistaken for stagnation and lethargy. The mission should be regularly reviewed, and its implementation needs to be observed. The suitable conditions need to be created for ambitious and bold attitudes, which cannot be justified with momentary goals. It encourages people to continue their work, although the efforts may not bring quick results and sometimes may differ from dominating tendencies. Even in continuous strategies, mastery is employed to create new approaches and attitudes in order to fulfil the mission more effectively.

The sustainable leader, committed to the organizational or common mission, takes charge and advocates the change by constantly seeking evolution in order to fulfil the mastery needs of the mission. The combination of the mission and mastery is the basis for the transformation, which begins the personal change.

The second phase of the change cycle is personal change. The reach for the mission mastery stimulates the change, which includes new ideas, new products, new services, and new organizations, i.e., all new ways to do work. Nonetheless, it must be stressed that it is not the company, structures, or processes which are changing, but the people and, first, the leader, who promotes the change. Having chosen a new path, the sustainable leader has to be the first to follow it. Sustainable leadership requires a continuous personal balance between the renewal of the existent things and the investment into new methods and competencies. The experience is important if it does not prevent the search for better examples. In any case, the mission demands a constant learning attitude. In this path of learning, the sustainable leader has to be courageous enough to explore new ways and to overcome any related doubts. Together with courage, the sustainable leader should demonstrate humility, which is needed in any learning process. He or she needs to learn to listen, to accept help, to test new methods, to try again despite disappointment, if the attempt was unsuccessful, etc.

The sustainable leader does not hold onto his or her assumptions (having in mind that the change is oriented towards the organizational mission and not him or her personally) and is not afraid to lose authority by admitting his or her mistakes. On the contrary, the sustainable leader grounds his or her prestige on constant efforts to improve oneself in any required way in order to fulfil the mission. In this learning process, it is beneficial for the leader to always have his or her own improvement plan with particular means, directed towards the areas he or she has to improve on in order to fulfil the mission more effectively. The personal improvement plan includes improvement goals which encourage personal change. Sustainable leaders have to openly share their personal improvement plan with their subordinates, encouraging them to give feedback and support.

The second phase of the change cycle is advocating followers to change. Despite it being a long process, once the personal change begins, the sustainable leader has the necessary power to attempt to change the subordinates. The mastery of the sustainable leader is not achieved with the capacity of one person. The leader can attain it only with the help of other people. Therefore, the change must be advocated from the top down. Many leaders try to force their subordinates to change; however, usually those changes are false. People adapt to new requirements due to obedience or due to their trust in the leader. Nevertheless, they do not consider more profound reasons to change. In many cases, it happens because the leader stresses the results and does not emphasize the mission and its requirements. Furthermore, the leader does not pay enough attention to the employees' needs or difficulties they are faced with in performing the change. The crucial point in this phase is the subordinates' movement from adapting to the change to actually mastering it. This objective can be reached when they understand that the change is needed for the mission's mastery. The sustainable leader must demand changes and the employees' commitment to the mission should inspire the efforts and learning, which is required in order to fulfil the mission. As the sustainable leader has the plan for personal improvement, the subordinates must have it as well. If the employees seek to realize their personal improvement plan and they understand the mission, they also master the change. The explained change process is a coaching practice, which will also be discussed in this book. The sustainable leader must be the team's coach, a close person with whom team members can discuss their problems and needs, which occur when advocating the mission-oriented change. Only then can sustainable leaders help their subordinates to actually master the changes required to fulfil the mission. Therefore, the leader must always find time for the coaching of subordinates.

The fourth phase of the change cycle is the creation of new leaders. The change is completed when the leader's followers become the advocates of that change, i.e., leaders in their turn. It is the longest phase in the change cycle. The subordinate becomes the leader and the leader becomes the chief leader or the sustainable leader. The sustainable leader does not supervise "from above." He or she passes the leader's baton to the subordinates down the ladder till the organization's base.

In such a way, step by step, the process of sustainable leadership creates new leaders. However, it has been observed that the leadership spreads downwards only when the subordinates move from the simple mastering of the change to advocating it. The essential factor in this process is the mission's division into common missions. The very organizational mission, based on the fundamental values of the organization, transforms the leader into a sustainable one. When leaders strengthen their personal commitment to the common mission on their level, the seeking of the mastery, the personal change, and the advocating of the change spreads down throughout the entire organization. Every leader seeks to acquire the mastery to fulfil the mission of his or her level and advocates the needed change. Every leader aims to achieve mastery by promoting the required changes to ensure it. Thus, by encouraging the change, the subordinates begin to see the mission with the leader's eyes. They discover new challenges and new ideas to realize the mission in more effective ways. A new change cycle starts in the never-ending search for mastery. The successful implementation of the change cycle mostly depends on the sustainable leader's rational intelligence. Nonetheless, physical, emotional, and spiritual intelligences are also important, as they are in the other phases.

The model of sustainable leadership requires a special type of leadership, i.e., sustainable leadership, in order to ensure the cultural change in the organization (Edmonds 2014). At the same time, this process facilitates the emergence of new leaders. Leadership is always the realization of oneself and the result of one's victories and losses, which are accepted nobly and with determination to learn. Therefore, the implementation of the sustainable leadership model requires support, provided by a serious and consistent program of sustainable leadership learning, the so-called coaching.

People and circumstances are always changing; thus, the process must be strengthened. However, sustainable leadership, implemented according to the presented model, can surely bring significant change in the organizational culture and allow the organization to fulfil its mission to transform into a sustainable organization, whose main values include environment protection, preservation of natural resources, and social obligations to the employees and the society.

## 4.6 SUSTAINABLE LEADERSHIP AND ENTREPRENEURSHIP

There are many types of entrepreneurs and the concept of "entrepreneur" is very flexible. The word entrepreneur results from the French *entreprendre*, meaning "to take in between," or "to undertake." English doesn't have its own word for entrepreneur. We use the French word in English because the proper word for entrepreneur is now used by another profession (someone who undertakes, a word used by the original theorists of entrepreneurship).

Presently there are cross-cultural entrepreneurs, mediapreneurs, end-poverty entrepreneurs, transparency-and-fairness entrepreneurs, social entrepreneurs, social-privatization entrepreneurs, world-citizen entrepreneurs, intrapreneurs, knowledge-collaboration entrepreneurs, cultural entrepreneurs, and biodiversity entrepreneurs. However, the main three are the most commonly used and cited literature:

1. Business entrepreneurs. They seek development and incomes within the business world. They are sustained innovators and are permanently trying to capture larger market shares from a competitive marketplace. Very often these individuals are sustainable leaders with all the qualities of a sustainable leader. They are original individualists who generate one project after another and one innovation after another.

2. Social entrepreneurs. They have several of the same personality features as business entrepreneurs, but they are driven by a task and seek to find advanced ways to solve problems that are not being or cannot be spoken by either the private or the public sector (Tyson 2004).
3. Small-business owners. They may once have apprehended an opportunity like an entrepreneur but then they rest on their laurels because they or the opportunity—or both—do not continue to have the attributes that make it entrepreneurial. The business may never produce on a large scale and the business owner may desire a more unchanging and less destructive approach to running their business. Many small-business owners often like stable sales, profits, and modest growth, and want to keep the business at a size they can personally manage and control. Often, they are called small-business managers, but not entrepreneurs.

Both business and social entrepreneurs look for innovation and development, they succeed in both small creativities and large organizations, and they have a mind-set that splits them from the rest of the population. Small-business owners would fairly exploit existing stable opportunities and optimize supply and demand in recognized markets.

It is important to stress that the traits characteristic of entrepreneurs is usually reflected in an entrepreneurial organization. In order to better understand the traits of entrepreneurs, Table 4.3 shows different definitions of the entrepreneur personality as suggested by various authors.

It can be noticed that most authors consider an entrepreneur to be a person who is capable of being a creator, observes all the possibilities surrounding him, and reacts to changes in the environment. An entrepreneur is focused and always knows what path to choose.

After analyzing the given definitions, the following features of the entrepreneur can be distinguished:

- creativity and innovation;
- initiative and active attitude;
- the tendency to find radical and unconventional solutions;
- use of opportunities;
- risk aversion;
- positive attitude to surprises;
- the ability to "motivate" your idea to others.

An entrepreneur initiates innovation and aims to introduce more innovation in practice, although he understands that innovation is associated with high risk, as well as the recognition that innovation does not always create value. An entrepreneur perceives innovation as a kind of performance development model, while taking risks if innovation does not generate the expected benefits. On the other hand, it is the initiation and implementation of innovations that can "unleash" the majority of the static, which at the time generates what has already been done.

**TABLE 4.3**
**Definitions of Entrepreneur**

| Author (year) | Definitions of entrepreneur |
|---|---|
| Schumpeter (1934) | Entrepreneurs are called independent business owners or players and all those who realize the "new combinations" creation functions, no matter what position they occupy at work (Hagedoorn 1996). |
| Shapero (1975) | Entrepreneur—a person who takes the initiative and creates social–economic mechanisms to realize it (Lucas, Cooper, and MacFarlane 2008). |
| Kanter (1983) | Entrepreneur and entrepreneurial organizations always act on their own within their competence, focusing their attention and resources not on what they already know, but on what they do not yet know. They value themselves not according to past standards, but to future visions. |
| Timmons (1989) | An entrepreneurs can be defined as someone who is able to see opportunities where others see only chaos (Kirby 2002). |
| Turner, and Pennington (2015) | Entrepreneur—it is a creative innovator who, acting on his own initiative, looking for opportunities, exploits them as much as possible, assumes the risk, and vigorously seeks a meaningful result. Money or profitability is not its purpose. |
| Drucker (2007) | Entrepreneur—a person who is constantly looking for change, responds to it and exploits it, caused by chance. The public and non-governmental sectors social entrepreneur solves problems and creates public space to allow citizens themselves to act and change their living conditions. |
| Židonis (2008) | In modern Western society, the global entrepreneur is considered to be a creator, an inventor, even a rebel, generating new products and implementing new services, thus seeking to defend "poor consumers" on the big monopolies oppression. |

*Source:* created by authors.

According to Lessem (1986), entrepreneurs differ in character traits (see Table 4.4). In total, he distinguishes seven types of entrepreneurs, the features of which determine the dominant personality type. One example could be the type of innovator whose predominant trait feature is creativity potential, based on the expression of imagination. Developed imagination for this type of transponder is a source of inspiration, transformation that reads original ideas. According to Shane (2010), the process of implementing an entrepreneurial business is influenced by the identification of opportunities, the environment, and the psychological factors that determine the expression of the potential for reciprocity. Types of entrepreneurs are presented in Table 4.4.

Entrepreneurship is more than the plain creation of a business or a social enterprise. Entrepreneurship is a combined concept that permeates an individual's enterprise in an innovative style. It is this mind-set that has revolutionized the way business and social projects are showed at every level and in each country.

In scientific literature, the development of an entrepreneurial concept is analyzed in the context of classical and neo-classical theoretical insights. The French school (eighteenth century) is represented by such prominent theorists as Cantillon,

**TABLE 4.4**
**Types of Entrepreneurs According to Lessem**

| Type of entrepreneur | The dominant personality type | Characteristics features |
|---|---|---|
| Innovator | Imagination | Originality<br>Inspiration |
| New developer | Intuition | Evolution<br>Development<br>Combining |
| Sustainable leader | Authority | Leadership<br>Responsibility<br>Structuring<br>Control |
| The new entrepreneur | Will | Achievement<br>Possibility<br>Risk-taking tendency<br>Power |
| Inspired | Sociality | Informalism<br>Shared values<br>Community |
| Adventure seeker | Dexterity | Activity<br>Diligence |
| Agent of change | Flexibility | Ability to adapt<br>Compassion<br>Curiosity/willingness to know |

*Source:* created by authors based on Stripeikis (2007).

Turgot, Say, and Baudeau, who said that the entrepreneur as a business entity is exposed to risk in a business environment (Landström and Lohrke 2010). The risk in the economic system includes an uncertainty factor, where the consistent development of ancillary business in the long run cannot be fully anticipated. According to Cantillon, recyclers buy raw materials for the production process at a fixed price, but cannot predict the fixed price of the final product (Ekelund and Hébert 2013).

In a situation of uncertainty in the market situation, the success (profitability) of an entrepreneurial business rose from risk management through strategic decisions. Turgot, as in Cantillon's theoretical paradigm, points out that there are functional differences between the capitalist and the recorder in the development of commercial activity (Nisbet 1980). Jean-Baptiste Say has incorporated natural, human resources and capital into the theoretical paradigm of production and distribution, which are important for ensuring the productivity of the product manufacturing process. Say has functionally defined the interaction between recruits, theorists, and hired workers (Schoorl 2012). Opportunities for cooperation between these three stakeholders ensure a harmonious production process and promising opportunities for the development of an entrepreneurial business.

Theoretics are scientifically based ideas—the knowledge is given to the trainers. Meanwhile, the masters use the acquired knowledge to develop the stages of process alignment and organization of the production process. Meanwhile, the hired workers are responsible for carrying out the works assigned to them. According to Say's theoretical insights, it can be said that the production process involves certain specific stages in which three stakeholder groups play an important role: theorists, recruits, and workers. Different functions result in a consistent sequence of production cycles. The recruiter is identified as a production system coordinator who occupies an intermediate position between theorists and hired workers. Say, like Cantillon, argues that a recruiter cannot predict how personal decisions will affect business in the future, by developing commercial activities under dynamic market conditions (Prendergast 2011).

The level of risk tolerance depends on the personal qualities of the entrepreneur, which help to properly and rationally assess the current market situation, which is important for the identification of changes in consumer needs. Therefore, the development of products or services must be oriented to the needs of the market, which determine the relationship between supply and demand (Baron 2012; Barzdaite and Navickiene 2013). According to Baudeau, the recorder has the ability to reduce production costs by integrating innovation into the business process (Gallaher, Link, and Petrusa 2007). Innovation encourages the discovery of new tools and ways to increase the productivity of an entrepreneurial business and to increase the volume of commercial activity accordingly.

Smith, representing the British School (second half of the eighteenth century), identified the entrepreneur as an "adventurer" or "designer," who not only risked but focused his time and investment towards the goal (Casson 2003). An investor invests his or her own resources in risky activities in order to obtain the expected benefits (to increase the current capital value and profit). The British School spokesman, Mill, has identified the term of the trainer to be related to his ability to see prospects in the foresight. Meanwhile, Marchall noted that not all people engaged in commercial activities could be treated as recruits (Amatori and Colli 2013). An entrepreneur is engaged in innovative activities to lead the way in risk factors (Bolton and Thompson 2004; Busenitz and Lau 1997). According to Marchall, the recruiter is an innovator in the business world to discover new opportunities and create new products or services (Casrud and Brännback 2007). It can be argued that Marchall has linked an entrepreneurial business with the creation and implementation of innovative ideas, based on an entrepreneurial activity.

From the nineteenth century, the Austrian School's representatives Menger, Mises, Wieser, Hayek, and Kirzner also dealt with the role of an entrepreneur in business. According to Menger, the market is dynamic, so the main goal of the recorder is to constantly get information about the current market situation (Casson 2003). Knowledge of the market situation and its trends is important because it helps to objectively assess the economic environment and identify business opportunities in which areas sectors can successfully develop an entrepreneurial business. Knowledge helps decisions to be made that determine performance, so market analysis encourages consideration of various strategic options, taking into account the needs of the users. From the point of view of Mises, the recruiter is a "speculator" whose commercial purpose is profit (Schoorl 2012).

The American School spokesman Knight argues that a person engaging in an entrepreneurial activity is able to adapt to the conditions of market uncertainty (dynamics), assuming full responsibility for the company's activities, creating innovative ideas (Amatori and Colli 2013). It can be said that an entrepreneur is closely related to risk and his profitability depends on the skills of the entrepreneur and the knowledge of risk management.

Schumpeter, a spokesman for the Neoclassical Austrian School, sees the teacher as an innovator in the theoretical paradigm (late nineteenth century) (Gallaher, Link, and Petrusa 2007). An entrepreneur promotes market changes by integrating innovation into business processes. Schumpeter (Ekelund and Hébert 2013) distinguishes key business combinations, including:

- new products or their quality;
- new production methods;
- new market;
- new ways of using raw materials;
- new ways of industrial organization.

Neoclassical theoretical trends are represented by Weber, who emphasized in his theoretical paradigm that there is a strong interrelation between economics and religion (Landström and Lohrke 2010). Economic action, exercise, and expression are influenced by religion as a set of values and beliefs. Anthropologist Weber is associated with the role of a charismatic leader, an exclusive personality in economic activity, in theoretical insights (Amatori and Colli 2013).

There are a lot of descriptions of entrepreneurship in economics and management science. Some authors argue that entrepreneurship, as a process, is oriented towards creativity and innovation, while others believe that creating the value added is the main principle of entrepreneurship. But on one thing, different authors usually agree: entrepreneurship is the ability to see opportunities where others see obstacles. It is also accepted that entrepreneurship is an integral part of new value creation and non-traditional competition solutions.

The attitudes of different authors are presented in Table 4.5.

It has been noticed that, over time, the perception of entrepreneurship from different authors has remained largely unchanged. Entrepreneurship definitions, although associated with the times and social norms and regulations change, remained dynamic.

Summarizing the concepts presented by all the authors, it can be said that entrepreneurship is a type of activity or process that exploits all the opportunities that result in added value for others. Entrepreneurship has features such as proactivity, innovation, initiative, creativity, change management, and adaptability to a dynamic, fast-changing, and unpredictable modern environment.

To better know the relationship between sustainable leadership and entrepreneurship, it is significant to consider entrepreneurship theory development.

**TABLE 4.5**
**Definitions of Entrepreneurship**

| Author (year) | The concept of entrepreneurship |
|---|---|
| Hisrich (1986) | Entrepreneurship is a method of creating new value, with the necessary time and effort, taking on the potential financial, psychological, or social risks and generating money and benefits that are the main motive for all activities. |
| Timmons (1989) | Entrepreneurship is the capability to create something out of almost nothing. These are more initiation, making, pursuit than observation, analysis, and description (Kirby 2002). |
| Stevenson and Jarillo (1990) | Entrepreneurship is the process by which individuals seek the possibility to set up a business, even regardless of whether they have enough resources to accomplish this. |
| Kirby (2002) | An individual or organization-level response to the challenges of a rapidly changing environment. Entrepreneurship is a way of working with features such as observing opportunities, proactivity, and innovation. |
| Turner and Pennington (2015) | Entrepreneurship is voluntary cooperation, risk-taking, creation, implementation, the ability to raise and successfully implement innovative ideas that seek to maximize opportunities beyond existing models, structures, and resources. |
| Wicham (2006) | Entrepreneurship is defined as creating and managing a vision and passing it on to other people. It is taking the lead, motivating people, and preparing them for change. |
| Hisrich, Peters, and Shepherd (2006) | Entrepreneurship is a process of creating a new value for which the entrepreneur devotes the necessary time and effort to the potential financial, psychological, or social risk, and receives financial benefit and personal satisfaction. |
| Baldassarri and Saavala (2006) | Entrepreneurship is a person's ability to turn thought into action. It includes creativity, initiative, innovation, risk-taking, and the ability to plan and manage projects to achieve specific goals.<br>Broadly speaking, entrepreneurship is the mindset that can be adaptable to both business and personal life. |

*Source:* created by authors.

For a better understanding of the nature of entrepreneurship, it is important to reflect on some of its theory development. The study on entrepreneurship has grown intensely over the years and, as the field has developed, study methodology has advanced from empirical studies of entrepreneurship to more background and process-oriented study.

A theory of entrepreneurship is defined as a demonstrable and logically clear formulation of relationships or underlying principles that explain entrepreneurship. These principles forecast entrepreneurial activity or deliver normative guidance (Connelly, Ireland, Reutzel, and Coombs 2010). Entrepreneurship is interdisciplinary, which means combining fields and crossing boundaries between disciplines or schools of thought. It contains various approaches that can increase one's understanding of the field (Gartner 1990). Therefore, we need to know the diversity of theories as an appearance of entrepreneurial understanding. One way to inspect these theories is within a "schools of thought" method that divides entrepreneurship into detailed activities. These activities may be within a macro or a micro view (see Figure 4.4).

The macro view of entrepreneurship presents a board array of external factors, such as education, physical infrastructure, culture, and financing, that relate to achievement or disappointment in contemporary entrepreneurial schemes. These factors contain external processes that are sometimes outside the control of the separate entrepreneur, for they show a strong external locus of control point of view. The social and cultural school of thought contracts with external issues and surrounding situations and influences that touch a potential entrepreneur's lifestyle. The focus is on institutions, values, and social norms that, grouped together, form a socio-political environmental context that strongly affects the development of entrepreneurs (York and Venkataraman 2010). The financial/capital school of thought is based on the capital-seeking process—the search for seed and growth capital is the whole focus of this entrepreneurial importance. The displacement school of thought emphasizes the negative side of group phenomena, in which someone feels out of place. The ecological school of thought is based on the idea that all is related with all everywhere (green economics and ecological economics).

The micro view of entrepreneurship inspects the factors that are specific to entrepreneurship and are part of the internal locus of control. The potential entrepreneur has the skill, or regulator, to direct and regulate the outcome of each main influence in this view. Unlike the macro approach, which focuses on events from the outside looking in, the micro approach concentrates on specifics from the inside looking out.

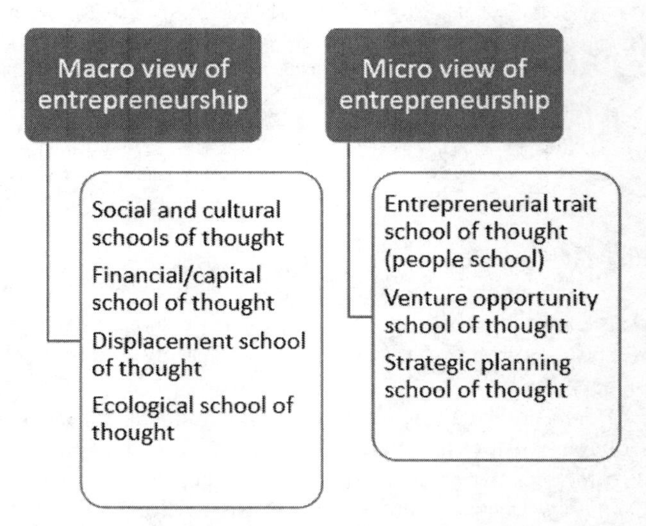


**FIGURE 4.4** Entrepreneurial schools of thought. (Source: created by authors.)

The reasons for the development of the concept of entrepreneurship in the organization, and the assumptions and consequences of the implementation in the organization are presented in Figure 4.5.

Globalization is accelerating changes that are also being driven by the development of information and communication technologies. Thanks to this, the awareness of the members of society increases their attitudes and interests, as well as the general public opinion. Changing public opinion allows new forms of public policy

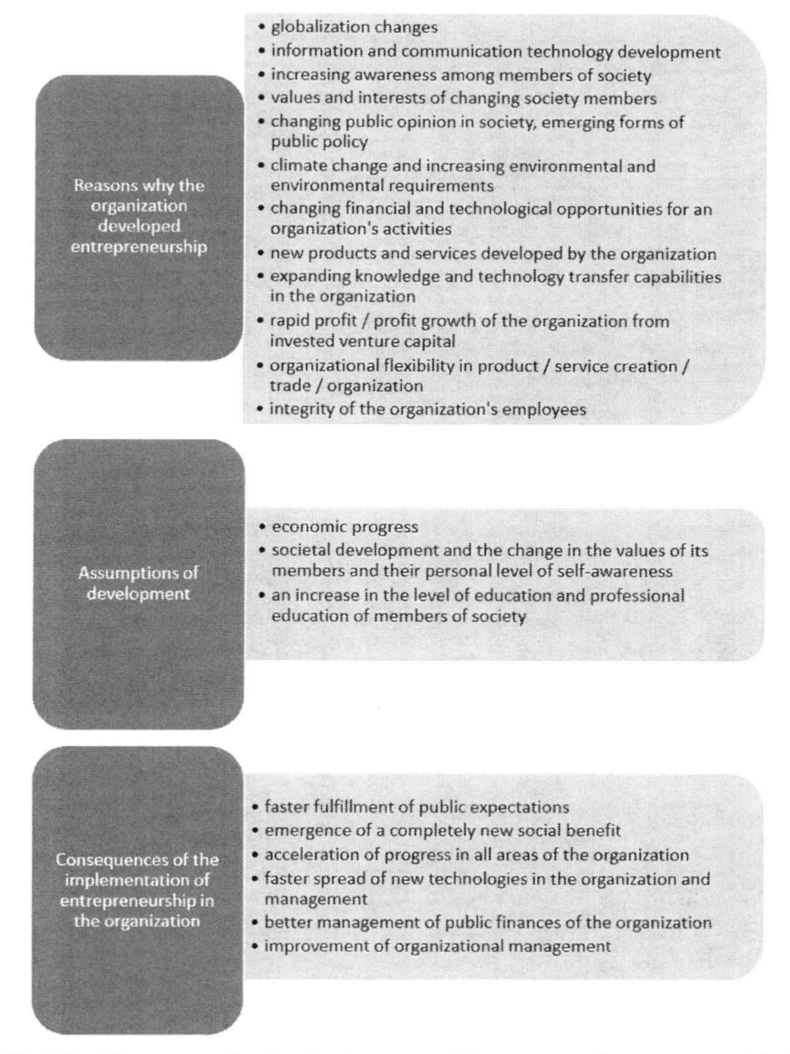


**FIGURE 4.5** The reasons for the development of the concept of entrepreneurship in the organization, and the assumptions and consequences of the implementation in the organization. (Source: created by authors.)

to emerge. Climate change and increasing environmental requirements are gaining increasing attention from the public and politicians, and through mutually agreed policy and business decisions. Changing financial and technological capabilities allow the organization to develop new products and services, and the expanding knowledge and technology transfer capabilities.

All this together enables the organization to invest capital riskily, quickly gain profits, and expect profit growth. The organization becomes more flexible in product/service creation/marketing/organization, and the new organization of work increases the integrity of the organization's employees. The reasons for this development in the organization are based on certain assumptions, economic progress, the evolution of society, the change in the values of its members, and the increase in their level of personal self-esteem and the general increase in the level of education and professional education of members of society. The reasons mentioned lead to the consequences of implementing an entrepreneurship in the organization, such as faster fulfillment of public expectations, the emergence of a completely new social benefit, acceleration of progress in various activities of the organization, faster ramp-up of new technologies and management, better management of public finances, and improved management in the organization.

Although every company wants to be innovative, flexible, and creative, remaining entrepreneurial while making the transition to some of the more administrative traits of managers is vital to the successful growth of a venture. Entrepreneurs work with an effectual logic, creating, crafting, and adapting "means" to meet imagined new "ends." While managers use causal reasoning working with given "means" to achieve given "ends," the strategist adopts creative causal reasoning, by shaping given "means" and creating new "means" to meet given "ends" (Sarasvathy 2005).

To illustrate this point, we can develop the argument by adapting a recognized corporate strategy tool, called the Ansoff product–market matrix (Ansoff 1957), which plots the level of technology newness embedded in the product along the x axis and the level of market newness along the y axis (see Figure 4.6). The effectual logic has a place in Ansoff's Diversification quadrant. Here we are constrained neither by known products nor by known markets. Imagination, vision, confidence to sell, ambition, and so on will be necessary to succeed. This is the world of entrepreneurial reasoning and effectual logic, imagining new possibilities (the ends) and arranging the resources (the means) to achieve them.

Entrepreneurship of various types can originate in any quadrant, but over time any and every business will need, at different points in time, the service of these different forms of reasoning and logic. Managing the journey of the firm as it encounters the various quadrants will require calling upon different rational and logic approaches. Each point of view—entrepreneurial, strategic, and managerial—accounts for different considerations that need to be balanced across time if effective growth is going to be achieved. Entrepreneurial leadership may be the most critical element in the management of high-growth ventures. Relations such as "visionary" and "strategic" have been used when describing different types of leaders. The strategic leadership is the most actual for the growing organization. Researchers have recognized some of the most important ideas in effective strategic leadership. This kind of leadership can be categorized as entrepreneurial (sustainable) leadership,

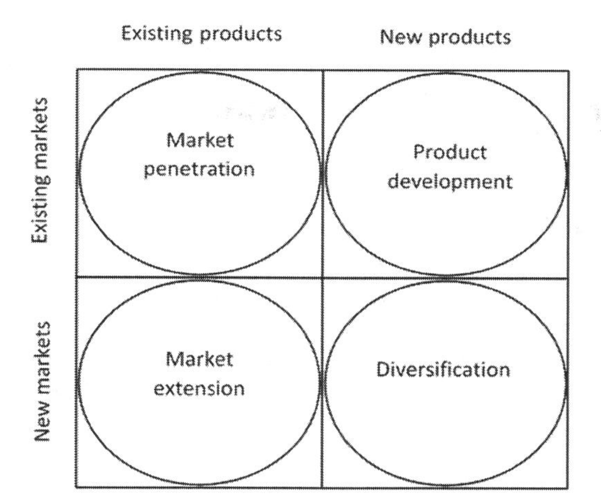


**FIGURE 4.6** The product–market matrix. (Source: created by authors.)

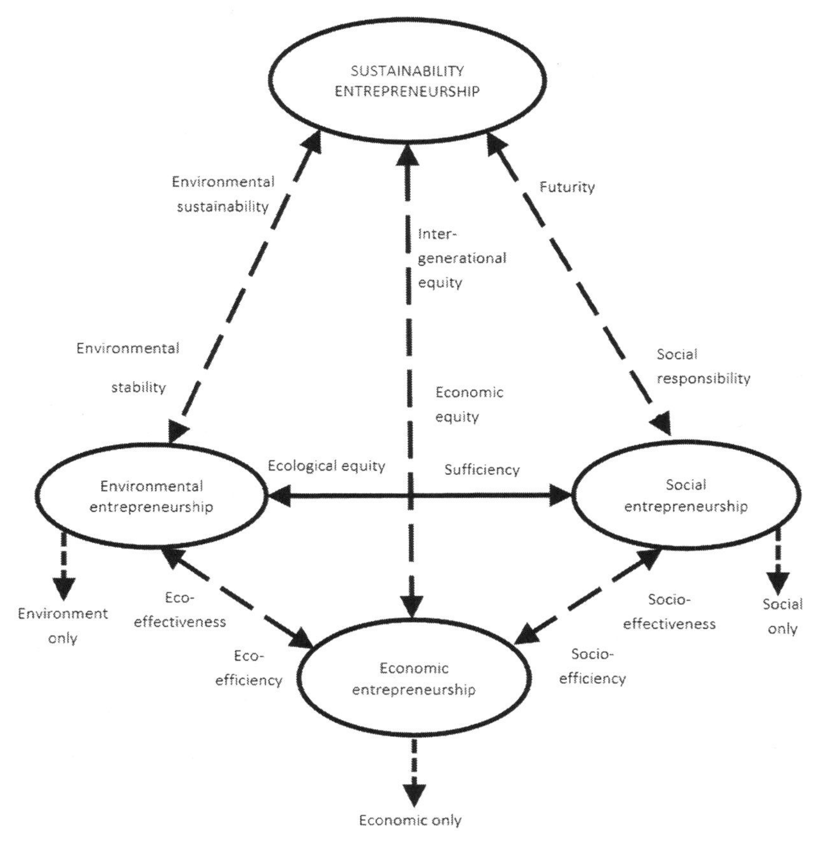


**FIGURE 4.7** The three branches of sustainability entrepreneurship. (Source: created by authors based on Tilly and Young (2009)).

which occurs when an entrepreneur manages a fast-paced, growth-oriented company (Rowe 2001; Ireland and Hitt 1999; Hitt, Ireland, Camp, and Sexton 2001; Alvarez, Ireland, and Reuer 2006).

Entrepreneurs have the ideal characteristics required to experiment, take risks, and put into practice these elements of the model (see Figure 4.7) and move toward sustainability entrepreneurship (Tilly and Young 2009).

Understanding sustainability in the context of entrepreneurship is becoming more sophisticated. It is important to focus on how entrepreneurs can use sustainability concepts to guide long-term development of competitive advantage. Integrating environmental and social goals into corporate objectives is becoming an imperative for all. The model referred to in Figure 4.7 illustrates the three branches of sustainability entrepreneurship and argues that neither social nor environmental forms of entrepreneurship are sufficient to qualify as sustainability entrepreneurship.

# 5 Sustainable Leadership Practices

## 5.1 COACHING FOR THE DEVELOPMENT OF SUSTAINABLE LEADERSHIP

In practice, educational management or coaching is widely used to develop sustainable leaders. In literature, coaching is divided into business coaching, life purpose coaching, cognitive coaching, transformational coaching, and relationship coaching. Different models of coaching are presented, e.g., GROW model, T model, etc. Each of these models has its own specifics.

It can be stated that coaching methodology covers several stages: confrontation and perception of the problem; determination of the objective; perception of reality (real situation); provision of alternative solutions; decision-making, clarity, the anticipation of support, and determination of operating strategy. Coaching provides an opportunity to form a purposeful work of the coach and client, where an independent person or a new leader is being developed who can realize his potential, be able to predict and make decisions, and be responsible for possible consequences. In coaching methodology, an important role is given to a coach, whose objective is to adjust the available resources of a person with a specific situation in order to achieve maximum results. When submitting questions, coaches encourage people to discover the necessary skills and effective powers in order to develop them and bring about changes and results. This process is guided by the principle that clients are the experts of their own problems; therefore, they know themselves and their life better. While creating an atmosphere of cooperation and trust, the coach motivates his client to find and make a decision as well as to implement it.

Lithuanian scientists use the term educational management to describe coaching. Coaching is an educational method and technique of training an individual or a group of people to achieve a certain objective or to acquire new skills (Hudson 1999; Ting and Scisco 2006). This is a methodology for increasing an individual's personal "effectiveness" in order to promote changes and achieve greater results in different areas. Educational management theory is based on the belief that every individual has sufficient innate abilities to achieve his objectives. The form is a purposeful formulation of questions (performed by a manager) and a search for answers (performed by a guided person). The coach first encourages people to discover necessary skills within themselves and inspires them to develop these skills to achieve changes and results. This technique is widely employed in organizations in order to achieve more effective personal improvement, learning, communication, and competence of employees. Recently, coaching has become extremely popular and even fashionable. People are interested in its abilities to help them to find a way out of a tricky situation, both in personal and in professional life, and to make a right decision in order to achieve personal or occupational objectives.

As has been mentioned before, several types of coaching are distinguished. One of them is life purpose coaching, which is "in-depth" coaching: the practical application of coaching in personal life by understanding the internal factors, resources, and possibilities. During this coaching, the focus of learning is not on how to ask questions but on the deep understanding and awareness of yourself and others. This coaching concentrates on the relationship with oneself and others, feelings and emotions, inner energy, as well as state, attitudes, and beliefs. Meanwhile, individual coaching focuses on particular wishes, objectives, perspectives, and thinking expressed by the client. Executive coaching aims to develop entrepreneurship.

The term coaching originates from the English language (a coach is a trainer, instructor, or tutor). One of the most commonly used definitions is that of a trainer. In this book, we use executive coaching and coach terms (Weiss 2004).

In other countries, the term executive coaching is widespread; however, it reached Lithuania only about ten years ago. When a new profession emerged (coach), there was an issue with definitions, exercises, and relations to other sciences. The unifying feature of coaching and similar occupations (e.g., management, counselling, and teaching) is the learning process.

Coaching is based on a very important assumption that people can achieve all of their dreams and objectives if they believe in them, live for them, and sincerely desire them. Thus, the coach who works with this method does not explain and advise but listens and asks questions of people who set their own objectives and find the optimal ways to achieve them (Donnison 2008). Clients who are involved in coaching see their activities, life, and themselves in a new light, as during the consultation they have to answer these questions: "Who am I?," "Why am I like that?," "Where am I going?," "What do I need to do to make the chosen way of solving the problem successful for a new stage of my life?"

The coach aims to make the client aware of the problem, and then analyzes and solves it, i.e., the coaching methodology is based on trust and understanding; it is directed towards the perception of the client's personal desires, and the revelation of specific objectives important to the client. This coaching technique helps to widen the boundaries of human reasoning; to understand themselves, their wishes, and objectives; to understand other people and their behavior. Therefore, managers and leaders who work by applying this technique are able to find more effective methods of educational development to motivate employees and encourage their loyalty, to use employees' potential to achieve the company's objectives, to effectively manage changes, and to motivate employees to enjoy their work and life. Executive coaching is very important for managers-leaders as it helps to ease employees' daily work processes and to manage employees by listening rather than administrating. If traditional management methods primarily focus on monitoring, provision of feedback, and information transfer, executive coaching is a less formal process, in which the manager-leader, by applying coaching methods and practices, listens, supports, and advises his employees. In addition, he provides suggestions, facilitates their personal and collective efforts, and creates conditions under which employees find ways to improve work performance (Graham 2005). In order to successfully implement coaching, the manager-leader must have the following skills: listening (recognition of feelings and their disclosure); ability to provide feedback; and ability to agree on tasks.

Listening is the key skill of a manager-leader for successful coaching practices. It consists of these elements: active listening, open listening, and the potential self-expression of the speaker. Active listening means that the listening process requires a lot of effort; it also means an active focus on the employee as well as the clarification of what he says. This is the way of listening between the lines. A person who actively listens to the speaker does not interrupt him but paraphrases, i.e., differently formulates and summarizes the information or ideas provided by the speaker. Active listening is an effective measure of executive coaching as it shows that the manager-leader is actively listening to what the employee is saying, while the employee sees this active listening as the sincere desire of his manager-leader to understand what he says. This type of listening gives the opportunity for managers-leaders to correct misunderstandings and make sure that they correctly understand what the employee has said; it creates mutual sympathy and understanding. Open listening is listening without prejudice: refraining from judging and allowing people to express their thoughts. Managers-leaders should not assume that they know the answer before the person has described the problem or told the story. The best strategy for solving problems is to devote time to clarifying them and only then to proposing solutions. To lead while educating your team members means that the leader is not in a hurry to make a decision; he encourages employees to thoroughly consider the problem, even waiting for them to find out the essence of it and allowing them to come up with their own solution. Giving the speaker the opportunity for self-expression shows that the main part of listening should focus on the encouragement of the speaker to express his thoughts, feelings, and aspirations. Open-ended questions, which start with interrogative words "Why?," "How?," and "What?," help to do this. The purpose of these questions is to help the speaker to explain the problem or issue in more detail. Open-ended questions differ from closed-ended questions as the latter must be answered either by "yes" or "no."

Provision of feedback means a clear, sensitive, specific, and constructive reaction of the manager-leader to particular actions of team members and their consequences. Feedback does not mean clapping people on the shoulder and kindly smiling, as such behavior from the leader can make an impression of patronage and mean that the manager-leader is patronizing this person. On the contrary, feedback should be focused on the concentration of employees and the improvement of their performance.

The agreement on tasks is the most important aim of the leader. A leader must constantly ensure that employees have a clear understanding of the unit's activities, objectives, and tasks of individual employees. The manager-leader must help the staff to clarify the tasks and agree on them. The proper balance of employees' workload significantly contributes to the performance of their unit; thus, the leader (and all employees) must make sure that no employee is overloaded with work and unable to cope with it.

In leadership, coaching or executive coaching is characterized by the fact that in his activities the manager-leader is guided by the following principles: he is committed to the obligations to provide support for people; his relationships with his learners or followers are based on truth, openness, and trust; the learner (team member) is responsible for the results which he seeks; in the future, the learner can achieve significantly better results than those he can achieve now; the leader always remains attentive to the learner's thoughts and experiences; the leader knows that the learner can find perfect solutions and tries to help him as much as possible (Goldsmith and Lyons 2006).

Thus, coaching is an effective educational tool for developing the leader's followers and new leaders; this tool unlocks the human potential to take action and seek maximum results. An individual is not taught, but he is helped to understand how to learn by himself. This is a process primarily based on a partnership between the leader and follower, which, by provoking the learner's thinking and creativity, encourages him to make the most of his personal and professional potential. That is why the art of communication plays a key role in this process. When presenting proper questions, the development of the learner's desired scenarios begins. This requires the learner to name his desired future, to identify available resources, and to predict necessary actions. This technique is particularly beneficial for those who want to understand the changes that are happening around them as well as to recognize and identify which changes are the best.

During the conversation with the coach, learners are forced to look at everything from a different perspective, they start to think more broadly, more objectively, they better understand the mission of the organization, easily accept the values of the organization, and enable them in their activities. The coach enables a deeper look into the learner's viewpoint, attitudes, perception, and factors that limit them (Hart, Blattner, and Leipsic 2001). The manager-leader, by using coaching practices, must, first and foremost, feel responsible for training and educating each team member at the workplace. In order to achieve effective participation and involvement of each employee in the implementation of the mission and objectives of the organization in all situations which require decision-making, managers-leaders can use this effective method of team motivation. This method of management does not allow the manager-leader to personally make all decisions, boss around, and control employees. Although it has to be acknowledged that this method does not solve all the problems of effective management, it creates very favorable conditions for employees to develop and find the best solutions by themselves. Knowing their objectives and understanding how the company helps them to achieve these objectives, team members become loyal to the organization and leader (Hart et al. 2001). Offers made by competitors do not attract them. They are motivated by the objective itself, and not by the material goods that many companies can offer. First of all, they are highly motivated by the attention of the manager-leader and his considerable effort to implement not only the company's but also their personal objectives. Each manager must be able to make decisions, provide instructions, and require employees to achieve their objectives. However, there are situations when the professionalism of the manager-leader is better expressed by applying executive coaching.

The manager-leader manages his team members by applying a developing style, which is constituted of a conversation with a team member about personal and organizational objectives, tasks (personal and organizational) and forms for achieving them (development of skills, pursuit of specific knowledge, reduction of discomfort, ways of support), internal existing and mobilized resources, established tasks (plans), and agreements on control meetings for discussion of results. In the attentive environment, the ongoing dialogue between the manager-leader and a team member facilitates movement towards the most important objectives in which they are interested.

Work, during the practice of executive coaching, starts with the setting of personal objectives and then continues with looking for connections with the objectives of the organization. Later, measures are set to achieve these objectives; unused opportunities and resources are sought. In this process, the only responsibility of the manager-leader is to ask proper questions which would help employees to find optimal answers. Executive coaching can be based on either the results or observation (Kampa-Kokesch and Anderson 2001). When executive coaching is based on results, the manager-leader performs a comparative analysis of objectives and actual results which have been set for the subordinate. By analyzing the results of the employee, the manager tries to distinguish his skills and knowledge that the employee can use to achieve the intended objectives. The observational executive coaching enables the manager to monitor how the employee performs his work and encourages him to teach and give advice to his employee.

However, the main objective of executive coaching is to build a culture of trust, commitment, and personal responsibility within the organization, both internally and externally (e.g., with clients), leading to greater loyalty of employees, which increases the efficiency and performance of the organization. This leadership helps to develop such leadership qualities as the constant need for cognition, the constant need for social relationships, creation of personality, motivation, erudition, and competence. Open-minded people can always bring more benefits to the organization than those employees who look around themselves and see nothing more.

During this period, the objective of the assessment of the organization's human resources is very important. In the past, many organizations have forgotten their clients and now employees are often being forgotten (Kampa-Kokesch and Anderson 2001). It is frequently noticed that people at work are passive, irresponsible, and ineffective. They are ready to work as much as they are paid for by the organization in money or awarded by other material goods; thus, they need to be strictly controlled. This opinion is often only a preconceived assumption of the manager, which provokes employees to behave as expected. Managers-leaders take a completely different approach and see their team members as responsible employees with high potential.

Empowerment objectives are extremely important for the manager-leader. The word to empower means to transfer legitimate authority, to provide powers, and to give the right. Empowerment is one of the terms that is increasingly used in organizations, but is still unusual for managers and subordinates, as well as not always fully understood. Delegation which has become a habit gradually creates an atmosphere of empowerment in the work unit or team. In leadership, empowerment means a certain attitude towards work relations. Managers, subordinates, and the whole organization are treated as work partners. Empowerment is understood as the creation of a different, unconventional working environment and relationships. It is based on positive and optimistic assumptions about employees. These assumptions state that employees are creative and active, they care about their work, and they want their work to be important and meaningful. Essentially, the organization and the manager, as the person who represents it, need to exploit these aspirations of subordinates and enable employees to fulfil them. Empowerment is a phenomenon covering the whole work unit; however, it starts with the change of the manager's approach to his subordinates. The manager-leader is the initiator and supporter of this new approach to employees.

This objective of such transformation of management also ensures that executive coaching is linked to leadership development. Most of what is covered by the activities of executive coaching can be explained very precisely and effectively by using the existing leadership theories.

Summarizing the results of the research, it can be said that executive coaching is a creative method that helps learners to change through the formation of new attitudes and behaviors. Learners are not subjected to external knowledge, but they are taught to learn, to make decisions, and to take responsibility for their actions, which is one of the essential assumptions for individual leadership. The efficiency of executive coaching depends on the emotional environment and relations between managers-leaders and their followers. The process of practical implementation of the executive coaching process is presented in Figure 5.1.

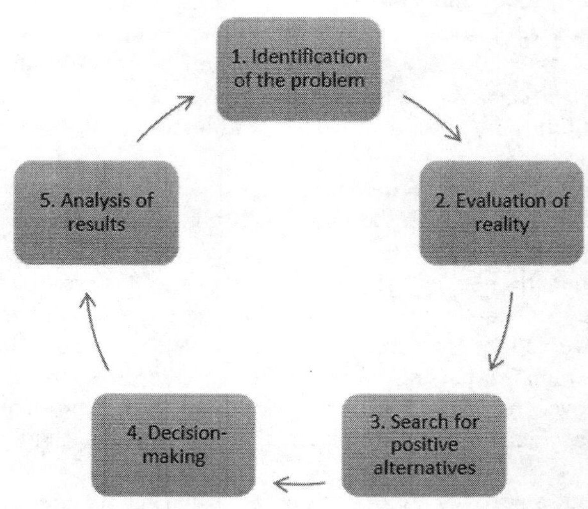


**FIGURE 5.1** The coaching process. (Source: created by authors.)

It is important to discuss and understand the specifics of each coaching stage.

1. The problem and its determination. This is the initial stage where the coach, together with the learner, indicates the problem. During the first conversation, the Circle of Opportunities can be used. The Circle of Opportunities method helps to define, specify, and narrow down the problem. The use of the Circle of Opportunities during the conversation is an effective measure to figure out what exactly is troubling the learner (Kampa-Kokesch and White 2002). In order to use this method, the coach needs to suggest that the learner draws the Circle of Opportunities (an example is presented in Figure 5.2). In the Circle of Opportunities, the learner should mark certain areas which are important in his work or life—for example, projects, duties, wage, financial resources, studies, family, relationships with friends, etc. In each area, it is suggested to mark points in the middle of the circle, depending on how much a person is satisfied with the quality of this area. The marking is done towards the center of the wheel in descending order, i.e., the closer to the center the point is, the less successful the person is in achieving the desired fulfilment in the marked area. These points are connected by distinguishing the main problem—as can be seen in Figure 5.2, in which the main problem is work. Once the problem is identified, it is possible to concentrate on the problematic area. It is believed that when working on one problem, the other problems will be touched on as well; thus, this will solve all the problems.

   In this stage of executive coaching, the perception of the problem, the perception of the real self, awareness, and other holistic leadership skills are developed.

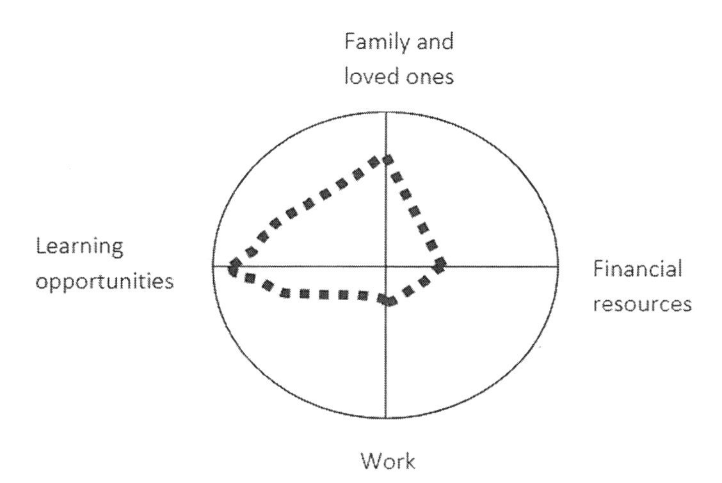


**FIGURE 5.2** Practical application of the Circle of Opportunities method. (Source: created by authors.)

2. Evaluation of the real situation. This is the second stage of the executive coaching process. During this stage, learners, with the help of the coach, carefully analyze their current life situation. All the facts related to the problem should be analyzed. At this stage, it is important to distinguish positive points from the problem. This can be done by using the Mirror Method. This method allows people to look at themselves from a different perspective, in this way allowing them to better understand and explain their actions, their life situations, etc. The second stage may take quite a long time. During this stage, facts can be revealed that would require them to return to the initial stage and to identify the problem once again. Moreover, during this stage, critical thinking, self-assessment, and diagnostic and other abilities of people can be revealed (Kampa-Kokesch 2003).
3. Evaluation of alternatives. This is one of the main stages where learners are taught to take full responsibility and make their own decisions, which are suitable for them. During this stage, the course of action on how to solve the problem is formed. The Lifeline Drawing method can be successfully used at this stage. This method enables learners to plan the main steps and to set the deadlines to implement them. At this stage, the coach mainly asks questions, allowing the learner to discover all the alternatives to the solution. Asking questions, as mentioned earlier, is the most important technique of coaching. Questions should be open-ended, short, and clear. The most productive ones are "What" and "How" questions because they are orientated towards the search for solutions. "Why" and "What for" questions should be avoided because they provoke criticism and force the learner to defend himself. During the evaluation stage of alternatives and opportunities, it is suggested to use one of the most important coaching principles, called the FCL rule: Fall in love, Change, Let go. First of all, fall in love; if you are not able to fall in love, then change it; if you are not able to change it, then let it go. This coaching principle helps learners to easier understand what they seek when solving their particular problem. It can be said that the introduction of this principle to learners will help them to take responsibility for their actions. At this stage of executive coaching, decision-making, planning, self-management, and other important leadership abilities are successfully developed.
4. Decision-making is the fourth stage. After discussing all decision alternatives, learners should decide which one is better for them and what kinds of actions need to be taken to implement it. At this stage, as in other stages, the coach should help, but not impose his decisions on learners, and he should not prompt direct answers. Hence, this is the commitment of the learner; it is his plan of action and his determination to implement his decisions. At this stage, it is advised to use the Empty Chair method. This method allows learners to make a decision thoughtfully, taking into account various factors and circumstances. At the stage of decision-making, responsibility, critical thinking, self-management, and other important leadership skills are developed.

5. The stage of result analysis. In fact, results are discussed all the time while the coach is working with the learner. The discussion of results, held after the execution of each decision, allows learners to think about where they are now and what kinds of problems they have faced while executing their plan of action, as well as allowing them to understand what else needs to be done in order to achieve the ultimate objective and to make them feel safe and fully-fledged in their environment. During the stage of result analysis, it is advised to use the Hot Seat method, which allows learners to analyze the problematic situation and its solution in more detail.

Hence, to summarize the stages of practical implementation of executive coaching, it can be said that executive coaching is an effective method to develop leadership competence; however, it requires competent coaches or managers-leaders who have mastered the coaching technique, time, and the desire and motivation of team members to join the coaching process. The effectiveness of executive coaching depends on both emotional and physical environmental conditions. A cozy environment where the conversation takes place and friendly behavior of the manager-leader will be prerequisites for successful coaching. Thus, coaching is a set of effective methods with an exceptional feature of bringing together and combining many well-known and used techniques, focused on the development and improvement of personal human skills into one system (Kampa-Kokesch and Anderson 2001). It is an active counseling process, disclosure of human potential, and the development of an individual by applying various coaching practices and strategies. First of all, abilities to make decisions independently, to take responsibility, to reflect on the performance, as well as other abilities important to the leadership competences are developed.

Therefore, coaching is directly related to leadership and the training of new leaders. Many objectives of executive coaching are consistent with the objectives of leaders which are set for team members and, in turn, develop leaders' competences. Moreover, it is also directly related to the motivation of employees. Most managers-leaders know that attention and respect for employees is one of the strongest motivational tools.

However, studies have revealed that employees who occupy important positions in the company often change jobs because they do not trust their manager, or their relationships are poor or undeveloped. For ordinary employees, a greater reward, career opportunities, training, or a flexible work schedule can be offered, while for the best and most important employees, this is already common practice. The same is offered to them by the competitors. The question arises: how to win the loyalty of the company's best employees and develop new sustainable leaders in the organization? First of all, such employees need regular attention from the manager-leader. After all, it is important for the best and most successful employees to find ways to increase their potential, make decisions more efficiently, and expand their capabilities. These objectives can be achieved if the manager and the team member work according to the executive coaching method. The objective of the manager-leader who works according to this method is to help important employees of the company to increase and exploit their potential while achieving their personal and the company's objectives. Executive coaching aims to help employees form such an approach, which would ensure success and promote effective behavior that improves the performance of both the employee and the organization.

Stevens (2008) carried out a study during which managers-leaders who apply the executive coaching methodology were interviewed. The aim of one of the questions, asked by the researcher, was to find out what, according to the managers, was the main objective of executive coaching. After systemizing the obtained results, the main objectives of executive coaching for managers-leaders were set: to gain a deeper, broader, and clearer understanding of the problems they face; to understand and cope with complex and ambiguous systems; to form and identify the culture and nature of the organization; to transform the organization into a sustainable organization by doing complex things in an appropriate manner.

There are four implementation phases of executive coaching in the organization. The sequence of executive coaching implementation in an organization is more orientated towards its objectives; therefore, this model is very important from a practical point of view. Depending on objectives and principles of executive coaching, the development of personal and social competences of the organization's employees varies; thus, in order to achieve the best results, it is often necessary to combine several different management techniques. The most common are the following: creative change technique; executive coaching technique based on neuro-linguistic programming; executive coaching technique based on psychosynthesis; executive coaching technique based on the Gestalt methodology; and solution-oriented technique.

The implementation process of executive coaching should cover four main phases: agreement, planning, implementation of actions, and briefing. To summarize the objectives, principles, and techniques of executive coaching, these key aspects can be distinguished: the main objectives of the executive coaching application are to figure out the organizational problems both within the organization and in its external complex environments; the organizational culture and climate should be formed adequately and the complex issues should be solved in an appropriate manner, i.e., by empowering employees in the process of organizational change in sustainable organization. It can be stated that the main objectives of executive coaching are similar to the organization's development and counselling objectives, while the key principles of executive coaching are defined as the development principles of management, leadership, and communication at all levels of the organization that impact the change of human resource characteristics of the organization. They are applied to develop personal and social competences at the individual, group, team, and organizational levels.

## 5.2 LEADERSHIP AWARDS IN LITHUANIA

In Lithuania, several competitions for the best manager (leader) are held annually. A competition which has the oldest and deep-rooted traditions is the Manager of the Year award. Manager of the Year is a project which was first organized in 2000 by the traditional weekly *Veidas*; it has been going on for 15 years now. The aim of this project is to distinguish the largest and most influential companies and their managers, to encourage positive attitudes towards business and to promote principles of modern management and leadership. This competition seeks not only to show the success of an individual person (manager) in business but also to develop public awareness of good management, leadership, and their importance in creating welfare of the country. The project is assisted by the Free Market Institute and the Confederation of Industrialists.

The culmination of the project is a solemn award ceremony held at the Presidential Palace usually at the end of February. The winners of the Manager of the Year competition organized by *Veidas* are as follows:

- In 2015, D. Misiūnas, CEO of *UAB Lietuvos energija*;
- In 2014, Andrius Rupšys, manager of *UAB Ruptela*;
- In 2013, Petras Masiulis, CEO of *Tele2*;
- In 2012, Algimantas Markauskas, CEO of Vilnius subdivision of *Thermo Fisher Scientific*;
- In 2011, Raimundas Petrauskas, CEO of *Schmitz Cargobull Baltic*;
- In 2010, Juozas Martikaitis, CEO of *Garlita*;
- In 2009, Sigitas Žvirblis, CEO of *Intersurgical*;
- In 2008, Linas Gipiškis, director of *Hnit-Baltic*;
- In 2007, Sauliu Jurgelėnas, director of the corporate group *Sanitas*;
- In 2006, Arvydas Paukštys, CEO of *Teltonika*;
- In 2005, Rimantas Kraujalis, chairman of the Board of *Eksma*;
- In 2004, Arvydas Juozeliūnas, group manager of *Nemuno banga*;
- In 2003, Tomas Juška, president of holding group *Libra*;
- In 2002, Darius Mockus, president of concern *MG Baltic*;
- In 2001, Julius Kvaraciejus, president of group *Pieno žvaigždės*;
- In 2000, Vladas Algirdas Bumelis, CEO of *Sicor Biotech*.

In addition to the awards of *Veidas*, there are other competitions too. Since 2011, *Verslo žinios* (VŽ) together with *Nordea Bank* have organized a project of Lithuanian business leaders for the top managers of Lithuanian businesses, and this project is crowned with the CEO of the Year awards. During the project, a leader who has created the most value for shareholders, employees, and society is selected, as well as leaders for five other nominations.

The CEO of the Year awards include four stages. Every year, VŽ makes a list of the top 1,000 largest companies in the country. In the first stage, companies that have finished the year profitably and have been headed by the same manager for at least two years are selected from this list. In the second stage, this list of 1,000 companies is given to the participating associations for evaluation. These associations recommend three candidates from the given list. Then the list of 16–48 proposed candidates is made. In the third stage, the list of candidates who have passed the second stage is given to the VŽ editorial board for evaluation. The editorial board evaluates the given candidates, their merits to shareholders, employees, and society, and then votes. After summing up the votes of the editorial board, 16 finalists are presented to the VŽ portal readers for their evaluation. The same 16 finalists are later assessed during the VŽ editorial board meeting, where the CEO of the Year and the winners of five other nominations are selected. The winners of the CEO of the Year and other nominations are announced and awarded during a special event dedicated to this project.

The following Lithuanian business leaders have already been awarded the CEO of the Year:

- In 2015, Algimantas Markauskas, CEO and vice president for the Baltic region of *Thermo Fisher Scientific Baltics*;
- In 2014, Raimundas Petrauskas, CEO of UAB *Schmitz Cargobull Baltic*;
- In 2013, Kęstutis Šerpytis, CEO and chairman of the board of AB *Lietuvos draudimas*;
- In 2012, Prof. Vladas Algirdas Bumelis, former CEO of *Sicor Biotech* and present chairman of the board of biotechnology company *Biotechpharma*;
- In 2011, Arūnas Šikšta, former manager of telecommunications AB *Teo LT*.

In Lithuania, there were more competitions of the best managers, such as a competition organized by Fastleader.com, which was held in 2007. Endrik Randoja, CEO of Fastleader.com, which is based in Estonia, has organized competitions for the best managers in the Baltic States since 2001. In 2007, the first awards for the best manager were organized in Lithuania. There were 500 executives of the largest companies and board members interested in it. The active voting revealed a great interest and determination to select a manager having the most leadership skills: candidates had been proposed by 300 employees. The main criteria were the growth rate and profit of the company's turnover, and how middle managers value their superiors. Jūratė Nedzinskienė, CEO of *SEB VB lizingas*, was announced as the winner during the conference "Leadership from the Heart."

After reviewing the results of awarded top managers-leaders, it can be stated that leadership is important for both manufacturing and service companies. Companies led by all selected executives have good financial performance, long-term development prospects, and well-established social responsibility principles. Many laureates are quite young. Young managers are characterized by the lack of attachment to old procedures as well as old and perhaps not always successful projects. For example, the previous manager, no matter what age, after starting some work and implementing new products, has believed in their future success, whereas the young manager is much bolder and more determined to end the unprofitable projects. A new manager always brings ideas and offers other methods of work organization. Many companies led by the best managers-leaders have implemented the Lean Management System, which is based directly on leadership principles.

Over the last decade, young leaders have been particularly impressed by the Lean Management System developed by the Japanese company *Toyota*. This system is similar to ISO 9000 principles and allows you to perfectly organize your business processes. It helps create greater value for the client while disposing smaller resources and increasing the company's competitiveness in the market. The core principles of Lean are continuous improvement and refusal of what hinders the effective organizational development. An organization that uses this method constantly strives for improvement and is not satisfied with what is already achieved. The Lean ideology can be compared to the aspirations of Japanese karate masters: they compete not for gold but for personal growth. This constant process of improvement requires a lot of effort from managers. If we compare the management of Lithuanian and foreign capital companies, it is obvious that foreign capital organizations pay more attention to the management practices in the areas of both personnel management and quality management, as well as in other areas. Lithuanian capital companies have traditionally been more oriented towards the conquest of the market and financial results. There manifests the effect of so-called "wrung sponge," as less attention is given to the development of management practices and the improvement of the organization.

On the other hand, currently, there is a growing number of Lithuanian capital organizations which, in this respect, are taking an example from foreign companies. Modern managers must be able to balance four priorities: work for the shareholder, employees, clients, and themselves. In a dynamic world full of stress and tension, they must remain both psychologically and physically robust and productive. Senior managers keep that balance. The head of the organization is usually valued not for personal results but for the results of his team. Sometimes a phenomenon can be noticed: a relatively primitive company works very profitably. It means that the company is capable of finding market-friendly opportunities on the market. However, its long-term prospects for survival in the market are low. In order to maintain and enhance its value, the organization and its managers must be characterized by comprehensive maturity of management practices; managers have to be real professionals and, first and foremost, leaders.

Lean-based companies strive to emphasize the importance of ideas and solutions provided by the employees to companies. While the Lean system helps companies get rid of unnecessary costs and become more effective, it is criticized for the stress it inflicts on employees through the implemented changes. According to Hofstede's (1980) research on cultural dimensions, Lithuanian society is hierarchical and paternalistic. That is, our leaders are afraid to ask subordinates' opinions, whereas the latter wait for instructions from the top, and avoid expressing dissatisfaction. Moreover, managers see suggestions for improvement as a personal criticism.

At the same time, the basis for the Lean system or a saving system is not only horizontal but also vertical cooperation. Cultural differences in companies are illustrated by the number of improvement proposals per employee. In Japanese factories, this number reaches 62 per employee; in Japanese factories in the US, this number is 1.4, and in American and European factories, there are 0.4 suggestions per employee. Successful implementation of the Lean system in Lithuania is often hampered by the lack of a basic environment (responsibility, accountability); however, socially responsible companies that provide reports on social progress, as a rule, use the Lean system.

Here it is also necessary to distinguish the concept of Lean leadership. Lean leadership is an essential element of the Lean system. First, it aims to involve every employee in taking the initiative to solve problems and improve their work. Second, it aims to ensure that the work of each employee would be focused on creating value for clients and the well-being of the organization. According to Fujio Cho, chairman of the board of *Toyota*, there are three keys to Lean leadership: go and see; ask why; show respect.

Executive managers have to spend a lot of time there, where value is created; they need to communicate with their subordinates and ask "why?" The "Why?" methodology or coaching questionnaires should be used every day. Managers-leaders, in particular, must respect their people. According to the concept of Lean leadership, managers-leaders do not wait for benefits by only delegating tasks to others; they also actively participate in the implementation of Lean.

The main driving forces of Lean leadership fit well with the driving forces of sustainable leadership: setting of objectives and mission (implementation of strategy; clear directions; explanation of objectives in daily activities; explanation of expectations); interconnection or cooperation (effective communication; team formation; motivation); empowerment or commitment (time management; prioritization; delegation, problem-solving; decision-making) and changes (personal changes, changes of team members, formation of new leaders). It can be said that Lean leadership is similar to sustainable leadership since it is based on related or even the same principles.

In the following section, we will discuss some of the winners of the competition organized by the weekly *Veidas*. We will focus on the features of their activities, character traits, and concepts of company management, expression of leadership and social responsibility, and the application of Lean and other advanced management systems, by taking into account aspects of the organization's culture and sustainable company development in the organizations they manage.

## 5.3 EXAMPLES OF GOOD LEADERSHIP PRACTICE IN LITHUANIA: GARLITA

In the competition of the Leader of the Year 2010 organized by *Veidas*, the manager of *Garlita*, Juozas Martikaitis, took first place. The commission had assessed the financial activity of companies, innovation, competitiveness, uniqueness of manufactured products and services, social activity, and social responsibility of the companies. This company and its manager have received a number of awards revealing the company's sustainability, innovation, high culture, and great social influence:

- In 2008, the company was awarded the gold medal as "Lithuanian Product of the Year 2008"—for innovative sweaters made from yarn enriched with the vitamin E.
- In 2009, the company was awarded the gold medal as "Lithuanian Product of the Year 2009"—for a sweater protecting against insects.
- In 2010, the company was awarded the gold medal as "Lithuanian Product of the Year 2010"—for a sweater resistant to cuts and sharp objects.
- In 2010, the company won the Innovation Prize and was announced as "Innovative Company."
- In 2012, the company received the award from the Lithuanian Confederation of Industrialists as "Successfully Working Company—2012."
- In 2012, the company was awarded the gold medal as "Lithuanian Product of the Year 2012"—for a multifunctional feminine knitwear jacket-coat.

- In 2013, the company received the award from the Ministry of Agriculture of the Republic of Lithuania for merits in business in the nomination "New Industrial Solutions."
- In 2014, Juozas Martikaitis received the Honor Award from Kaunas District Municipality for advanced economic decisions and charity.
- In 2014, *Garlita* received an award from the Kaunas Chamber of Commerce, Industry and Crafts for a significant contribution to the community.

Based on the 2014 *Garlita* "Environmental Report," the main results, innovation, and management and leadership practices of this socially responsible company are discussed further;

Located in the Kaunas district of Garliava, a Lithuanian knitwear company produces knitwear with nanoparticles that deter mosquitoes, protect against ultraviolet rays and bacteria, or are enriched with vitamin E. The company works with Austrian, Danish, and Swiss scientists, as well as those from Lithuanian, and with those who are creating such technologies. The company proves that even in textiles, one of the branches with the lowest value added, it is possible to create a very high value-added product. Creating such a unique sweater takes two years. The long-term manager and one of the owners, J. Martikaitis, is convinced that innovative products are the future of the company.

In 2010, the company started to produce sweaters resistant to cuts and sharp things and sweaters resistant to fire, and recently, they have created a new product with Danish scientists—a solar heat reflective sweater. When a person is wearing it, it does not heat up. The company produces various knitwear products, but the share of sales of sweaters with unique features is increasing. The company does not have any competitors producing this knitwear in Europe. After all, *Garlita* was the first in Europe to use raw materials with nanotechnology in the production of knitwear, allowing sweaters to be created which protect from ultraviolet rays and with other unique features. This direction of the company emerged a few years ago when the company had already earned the confidence of NATO troops and produced a military uniform for many countries.

The production of the first unique sweater using nanotechnology was very difficult—it was necessary to find scientists to find out how to "build in" the nanoparticles to the knitted fibers that persist both in the production process and when worn. A research institute in Austria, which specializes in military matters, agreed to try to create such technology. For the idea to pass to the production of the innovative sweater, it took one and a half to two years. Such sweaters are mostly purchased by the military of NATO countries. Sweaters with vitamin E are for soldiers serving in submarines, where they do not see the sun for half a year. Other countries are purchasing sweaters with ultraviolet protection for soldiers travelling in hot climate countries such as Iraq and Afghanistan. Cut-resistant clothes are acquired by the assault groups. Clothes with protection from insects are necessary for troops traveling to areas at risk of yellow fever and malaria. The company produced such outfits for the troops of South Africa and Nigeria. Cut and sharp object resistant sweaters were ordered by gemstone mines in Siberia. Moreover, the company produces non-flammable blankets and exports them to Scandinavian countries. When these countries banned smoking indoors, the demand for non-flammable blankets rapidly increased. *Garlita* also produces casual and women's clothing, high-quality clothes for famous European brands.

*Garlita* has earned an exceptional reputation, in particular due to the fact that the company's production is chosen by the majority of Lithuanian special services—police, customs, fire station; it is also trusted by foreign customers. The company collaborates with British, Swedish, Danish, Norwegian, Irish, German, French, and other companies. Production for special services is being purchased by the Swedish customs authority, the Norwegian army, the English emergency services and *Volvo* factories. The company, located in Garliava, has started to produce by special order upper knitwear for the famous Swedish sportswear company *Holebrook*.

About 95 percent of the company's output is exported to Western Europe, where sewed and knitted clothes by *Garlita* earned a reputation for being comfortable and high-quality products. The company's benefits include a favorable geographical location, excellent product quality, adherence to works deadlines, a skilled workforce, and the use of advanced technologies.

The second exclusivity is that the production of the company is maximally computerized, from all production procedures to accounting and the assurance of production security. In the production of new products, the environmental impact of the product is taken into account; therefore, the customer is recommended to choose a supplier whose raw materials are environmentally friendly.

The company is certified by ISO 9001, ISO 14001, and EMAS standards and has set quality and environmental policy objectives: to be a reliable partner of customers, respond to their legitimate needs, pay attention to expressed criticism, remove non-compliance and develop measures that prevent repetitive problems, and manage all the company's business processes effectively. Understanding the needs of stakeholders, the company has committed not to use hazardous materials that are harmful to the environment and human health in quantities that could endanger the surrounding population, nearby facilities, and nature. Moreover, *Garlita* seeks a clean and environmentally friendly, sustainable environment for future generations. The most important objective of the company is to create a clean, ecological work environment for each employee and form respective behavior of the workers across the whole company, create a high culture of the organization, and maintain a friendly relationship with the surrounding environment. *Garlita* is one of the companies which apply principles of sustainable development in their activities. Effective use of natural and human resources is the key principle of sustainable production. Therefore, the most important environmental aspects of the company are the reduction of industrial waste and the preservation of natural resources. For this purpose, the company installed an integrated management system in compliance with ISO 9001 and ISO 14001 standards, as well as implementing the requirements of the European Parliament and of the Council Regulation (EC) No 1221/2009.

The long-time manager of the company, J. Martikaitis, is an example of a real sustainable company leader. Through the efforts of the *Garlita* leader, integrated management systems have been implemented, and the culture of a sustainable company is being developed by educating responsible and committed employees.

Although *Garlita*'s range of products is not very large, environmental protection remains an essential engine of a sustainable business. The company is trying to implement such technologies which not only help to solve environmental problems but also cause the least possible damage to the whole ecosystem. By actively implementing the ideas of environmental protection and preservation of natural resources in production itself, the company aims to be one of the most advanced companies in its field in Europe.

Each year, the company sets objectives and tasks that are consistent with quality and environmental policies. The company's environmental policy must be known and respected by every employee of the company. The company's quality and environmental policies are published on their website. Appointed management representatives for quality and environmental protection ensure that the system works and brings real benefits in terms of development and environmental sustainability.

The environmental management system is based on managing the environmental aspects of the company's activities and products. The environmental aspects are managed in accordance with legal and other requirements of the company and in accordance with the principle of pollution prevention. Procedures have been developed and implemented to cover environmental aspects, activities, and emergency situations. All employees involved in the management system commit to comply with the environmental procedures. Employees are encouraged to make proposals, to improve the effectiveness of the environmental management system, and to participate in training as necessary.

The company periodically conducts internal audits of the integrated management system in order to continuously improve and prevent pollution. During the analysis performed by the management body, the information about the level of achievement, quality and environmental indicators, and other relevant issues which ensure continuous improvement of the environment and quality management system is being analyzed.

The company focuses on employee development, teaching, and training. Also, the manager of the company tries to ensure excellent working conditions and remuneration above the national average. Employees have good conditions for relaxation, as well as good quality food, and other important needs of the employees are taken care of; therefore, the turnover of employees in the company is very low and this is one of the most sought-after workplaces in Kaunas district.

Taking everything into consideration, it can be said that *Garlita* is a sustainable company in which a high culture of the organization with the help of sustainable leadership is being created. The company's commercial success is based on its innovative potential, applied innovative management techniques, environmental and social responsibility towards employees, and the community integration in the company's activities.

## 5.4 EXAMPLES OF GOOD LEADERSHIP PRACTICE IN LITHUANIA: SCHMITZ CARGOBULL BALTIC

Raimundas Petrauskas, CEO of *Schmitz Cargobull Baltic* (a German company that produces isothermal bodies, trailers, and semi-trailers), was chosen as the Manager of the Year 2011 by the commission organized by weekly *Veidas*. Also, Raimundas Petrauskas was chosen as the winner of the CEO of the Year 2014 competition organized by *Verslo žinios* and the bank *Nordea*. He was awarded the title of Chief Executive Officer of the Year at the Lithuanian Business Leadership Awards held at Vilnius Town Hall. In the first election stage, the citizen of Panevėžys had to overcome more than 670 competitors in the first stage of the competition and four of the strongest colleagues in the finals. For the first time in this competition, a person was chosen not from Vilnius. The manager of the manufacturing company which produces vehicles and refrigerators for supermarkets to transport goods states that it is the evaluation of the whole team and employees.

In addition, Raimundas Petrauskas is the only Lithuanian leader who won the two most important leadership competitions in Lithuania. He is very successful in managing a modern German capital factory in Panevėžys. The company has implemented an advanced Lean manufacturing system, which allows employees to work economically and wisely, eliminating unnecessary activities. When the company installed a Lean system, its productivity even surpassed colleagues from Germany. Similarly, a German factory produces four to six vehicles in eight hours and a factory in Panevėžys produces 16–18 in the same time. *Schmitz Cargobull Baltic* has a turnover of over 100 million euros. Lithuanians do not adopt the technology developed by the Germans but create new products themselves, and the engineers of the company are evaluated on the concern scale.

In an interview by the newspaper *Veidas*, Raimundas Petrauskas revealed what has made him most proud during his career as a manager. In the words of Petrauskas:

> this is a story of success: there was no lock on the door of the company that managed to create a good team, people have advanced workplaces, customers, markets. Everything has been created not by me, but by my team.

R. Petrauskas is a demanding manager who is able to inspire the team with his ideas. He is also constantly looking for new opportunities. R. Petrauskas believes that:

> if you rest on your laurels, tomorrow might not exist. We are looking, observing how the market changes because, ten years ago, the product life-cycle was long and now it is getting shorter—the product created last year does not necessarily bring success this year.

Acquaintances and friends emphasize that R. Petrauskas adheres to the given word, always tactful, modest, and polite, and usually communicates not only with his old friends, but with all the employees—shakes a hand and asks how things are both for a department manager and for a simple employee. R. Petrauskas' friends and former colleagues characterized him as a very dutiful person who works with Samogitian stubbornness from beginning to end and reaches what he has planned. These qualities have allowed Petrauskas to get to where he is now—to lead *Schmitz Cargobull Baltic*, one of the most advanced companies in Germany. During the years of his leadership, R. Petrauskas has created a modern company. He rescued the company from crisis for a second time in 2009 when turnover fell by 80 percent; he had to reduce the number of employees by half and had to live for almost a year without any income. In consultation with the employees, the working week was shortened, and it was agreed that when there are no orders the workers will learn, develop new products, improve processes, and prepare for better times. Such a strategy has proved its worth and the years 2011 and 2014 broke all the records in the company's history.

*Schmitz Cargobull Baltic* is one of the most desirable employers in Panevėžys. No wonder the average salary in the company exceeds 1,500 euros, and the manager of the company holds the position that, if the company is successful, it also must be felt by the employees. The manager of the company says:

> we have all grown together, learned and improved. Our team got involved with gifted engineers, financial specialists, managers, talented employees. Together we have provided a vision, created a strategy, we formulated the tasks and implemented them together. The people I work with are the greatest value.

R. Petrauskas has been managing the company for two decades since 1994. Petrauskas can be considered a good example of a sustainable leader as he complies with the position to work hand-in-hand with the team. The managing director can often be seen walking in the production workshops and talking with the employees, who are forced to look for solutions themselves. The manager of *Schmitz Cargobull Baltic* deliberately overturned the pyramid of the company's hierarchy and gave subordinates the opportunity to make their own decisions. This is an example of educational leadership and the empowerment of employees.

R. Petrauskas is certain that an employer who works a specific job knows it the best and sees the problems. According to him, no manager will make a better decision than the employees can offer themselves. The manager's concern is to find out what disturbs the employees from working effectively.

R. Petrauskas' determination to involve the team in decision-making at all levels has proved its worth. People who are not afraid to express their opinion make suggestions, feel needed and valued, and are more interested in the success of the company. This company focuses on employee development, training, and education. There is also applied training for managers and various coaching practices. Only a focused, qualified, and motivated team is capable of performing complex orders, responding flexibly, and quickly responding to the needs of customers. The manager of the company stresses that the company is glad to benefit its customers, as customers encourage the company to grow, develop new products, and install advanced production and management techniques and technologies.

The managing director of *Schmitz Cargobull Baltic* emphasizes that the company has also preserved some important traditions. That is, employee loyalty. A number of people have worked for 10–15 years in the company. Another parallel is creativity and the ability to create high-quality products. Not only is time and various cost control important, but also variables of management quality. The manager of the company understands that effective management and the application of the principles of leadership are not the same. Management is the pooling of information and control, while leadership is a reduction of uncertainties and deviations and employee empowerment. A manager who is a judge and controller is already a past phenomenon. The manager of the company emphasizes that "the future belongs to managers-leaders who teach and educate employees, show them an example and encourage them to act."

The company employs people of different generations. What once motivated one generation does not motivate the other. The director emphasizes that:

> the money is no longer the main means of motivation. Motivation is a psychological-emotional state which generates energy and makes the person act. In other words, if the employee has more energy than the manager has released explaining the task for him/her, then the objective is achieved—a person is motivated to work.

R. Petrauskas says that the present and the future must be based on the principles of sustainable leadership and develop new leaders-followers. In the modern company, there are universal truths—as in the laws of physics: it will inevitably have to move from managing to leadership, rather than looking back, will have to look forward, not avoid risk, and be able to manage it. The modern company moves from hierarchical management to partnership, does not indicate to employees but gives them the power not only to make decisions but also to take a certain responsibility for what they have decided. Such companies move from control to self-control as they have motivated people who are interested in the success of the company. "Those companies which are capable to change constantly create added value for partners, shareholders and employees, have more perspectives in the future. It looks so simple and obvious that sometimes it is hard to believe."

In summary, *Schmitz Cargobull Baltic* has achieved such good results as it is a high-cultural and sustainable company, although it is not a member of the Global Compact and has not implemented environmental or social management systems or standards; however, it is obvious that the manager of this company has all the characteristics regarding sustainable leadership competencies, and applies the principles of sustainable leadership and Lean production systems to create the culture of an organization, while ensuring the sustainable development of a company that implements the development of cost-effective production and environmentally friendly and socially committed employees, customers, and other stakeholders.

## 5.5 EXAMPLES OF GOOD LEADERSHIP PRACTICE IN LITHUANIA: RUPTELA

During the Manager of the Year election organized by the weekly newspaper *Veidas*, the founder of the company *Ruptela*, Andrius Rupšys, was elected Manager of the Year 2014. Moreover, the company, which creates and produces traffic monitoring systems, has been included in fastest growing company lists numerous times. The daily newspaper *Verslo žinios* has also chosen it as one of the *Gazelės*, the fastest growing small and medium-sized enterprises, and the company has been included in the *Deloitte* lists of the fastest growing Central European technology companies. It has also been included in the TOP1000 list of the biggest companies in the country.

The team of this company is very young—the average age is 28 years. In an interview with the weekly *Veidas*, the manager of the company, Andrius Rupšys, stressed the importance of the human resources office in the organization. The purpose of this office is to take care of the people and the organization's culture. This organization takes care of the employees, making sure that they are healthy, content with their work, and productive. The company also makes sure that the employees can eat, as there are kitchens on each floor of the organization. Once a week, the employees can order any products they like. A certain amount of money for the products is provided by the company. Due to the manager's observation that the employees like to order unhealthy food, lectures about healthy nutrition are organized.

In *Ruptela*, there is a sheet of paper on the door near the finance unit that details what each worker is responsible for. Nearby is a drawer where bills are left. The manager stressed that the employees of the finance unit discovered that these small details help them save around 20 percent of work time, as people distract them far less often. A children's room has been set up nearby as well. The employees' children do not spend the entire day there. However, if the parents need to stop by the workplace for a short time and have nobody and nowhere to leave the children, they have this space in the office.

The company also contains a prayer room. It is used by employees who practice Islam. In the beginning, they used to take the key for the archive room from the bookkeepers and pray there. With the intention of making the employees feel comfortable and less restricted, the manager of *Ruptela* established a prayer room. According to him, employing people from other countries has really paid off for the company. First, as the employees are able to speak with potential clients in their native language, there is a smaller chance of misunderstandings. Second, these employees know the culture and mentality of their home countries very well. Therefore, it is far easier for them to establish a connection with their compatriots than for a Lithuanian who has learned a certain language.

The company also boasts a sports room for balance exercises. A lecturer was invited to teach the employees. The office of the company contains a lot of vacant working space because the office was designed to be bigger than the company requires at the moment. The fast-growing company does not intend to slow down, thus the empty spaces are designated for future employees.

The implementation of the Lean methodology in *Ruptela* began before 2010. Consultants were hired. This method was implemented in small steps. The first step was the organization of daily meetings. The manager of the organization stressed that the company lacked meeting organization culture and Lean was very helpful, as the meetings are now short, structured, proceed quickly, and leave no time for long discussions and arguments.

The company also has an informal management board. For now, *Ruptela* does not have any sort of formal management. The manager of the company says that he had thought about creating it. However, the fast-growing and ever-changing company has always had more pressing tasks at hand. *Ruptela* has implemented a practice of leaders reporting to employees quarterly. This has proved to be very effective.

The manager of *Ruptela* possesses competencies expected of a sustainable leader and applies advanced management methods. A. Rupšys first asked himself some questions: Where is the company headed? What is its goal and what is its mission? When there is an understanding about where the company is headed, it is easier to see the bigger picture, to know which goals to prioritize, and for the employees to understand how to implement the goals better when doing their jobs. During conversations with the employees, the manager learned that they are very interested in what the other departments are doing, what is being planned, and what the company will look like after a year or two. At first, the manager prepared the presentations for the employees himself. However, with the company expanding, it became tricky for the manager to report to all the departments by himself. Then the idea occurred that this should be a job for the leaders of each of the departments. Thus, the tradition to meet quarterly for at least 2–3 hours, to discuss and answer everybody's questions, came to be. This practice helps increase employee involvement and responsibility, as well as allowing the leaders and employees to feel like masters of the goals.

As a true sustainable leader, the manager of *Ruptela* stresses all the time that it is very important for him to listen to what the employees are saying. That is how important changes happen in the company. A. Rupšys is certain that a functional leader or manager should first of all be a leader, but also a helper, educator, and teacher to his employee. That is why company leaders at various levels have to be accountable in front of the employees.

The company regularly holds training for the leaders. As the company grew, many of the leaders came from within the company itself. Some of them were hired from the outside. There are some who lack leadership experience. The manager of the company considers leader training and educating his duty. A. Rupšys admits that the constant changes and growth within the company have not been easy for him either. According to him, a couple of years ago, he thought that he knew a lot and considered himself an expert in many fields. A shift occurred when the company needed more high-level managers-leaders. Then he clearly saw that both the leaders that evolved in the company and those who came from outside often knew their respective fields far better than he did. In some cases, the entire work experience of the manager was shorter than a functional leader's period of employment in a certain field. The owner of *Ruptela* says that he understood that he must grow further as a leader, trust his employees, enable them, create proper conditions for them to work in, and lead them towards achieving their common goals. Therefore, the leader training program also includes a coaching program. A. Rupšys offers this explanation for his decision: "If a person has the necessary competencies, knows what to do and can do it, he only needs some help to find the answers."

A. Rupšys, first of all, feels responsible for the strategic goals of the company. These include turnover, client satisfaction, and involvement of the employees. The manager admits that the company is constantly changing. He says: "Change is not an easy thing, but that is the only way to live." According to him, the previous work was more focused on the short-term perspective. However, he noticed, that the decisions made now will determine what *Ruptela* will look like in ten or more years. A long-term strategy was formulated along with a long-term goal—to become exceptional in their sector within a decade; to offer the client unique value with their products and services. The goal of *Ruptela* is to become a global company. At this time, it is an international company and not global yet. The company needs to grow immensely and get a slice of the global market; therefore, it is necessary to work with much larger clients and, for that, it is imperative to increase their mastery in all fields.

All in all, it can be stated that *Ruptela* is a sustainable company. Although it is not part of the Global Compact and it has not integrated environmental or social management systems or standards, this company boasts a really high level of social commitment to its employees, applies advanced management methods, including Lean and the principles of sustainable leadership, and has created a strong organizational culture. The manager of the company possesses the qualities of a sustainable leader and all the leaders in the company are educated through coaching to become sustainable leaders. This company has achieved good economical results due first to their chosen path of sustainable development, and with the help of sustainable leadership, the company is being successfully transformed into a sustainable organization.

# 6 Sustainable Innovations

## 6.1 THE CONCEPT AND THE MEANING OF INNOVATIONS

Scientific literature offers no unified definition of innovations. Different authors have their own approach to this term. One of the first scholars to use the term in the scientific context was J. A. Schumpeter (1934). He claimed that innovations are a self-contained force which stimulates the economic growth and the economic cycles. The economic development is determined by a continuous emergence of new combinations (innovations) which economically have more prospects than previously used means to accomplish certain things.

Later, the number and the variety of definitions of innovations increased. Some of them describe innovations in a more detailed manner, others are more abstract. Nonetheless, many of the foreign authors associate innovations with a certain process, just describing it differently. According to Lundvall (1992), innovation can be defined as a process of continuous search and investigation, the result of which may include new processes, products, organizational forms, or markets. Innovation can also be described as a learning process which may arise from different factors: from learning by consuming, sharing and acting, as well as from the external and internal sources of knowledge and from the employment of the company's capabilities (Shenhar and Dvir 2007). Fagerberg, Srholec and Verspagen (2009) state that innovation is a process, during which a possibility is converted into new ideas, and during which those ideas are widely applied in practice. Akram, Siddiqui, Nawaz, Ghauri, and Cheema (2011) describe innovation in a similar way. They claim that innovations are a creative process, during which new knowledge is implemented in order to design products or services suitable for the market.

Various Lithuanian authors also propose different definitions of innovations. One of them is as follows: "innovation is the technological implementation or (and) development of new products and processes and it almost always involves obligatory technological modifications in companies" (Varkulevicius and Naudzius, 2005). On the other hand, Jakubavicius, Strazdas, and Gecas (2003) argue that innovation is a functional and advanced novelty which, in essence, is oriented towards the shift from old to new. Vijeikis (2011) offers a slightly different concept of innovations. According to the scholar, in a broad sense, innovations are a creative activity, which transforms the scientific knowledge into new or improved products that are beneficial to the business and society.

In their recent study, Rajapathirana and Hui (2018) put primary emphasis on the connections that innovations have with open new technologies and open resources and stress the importance of open innovations.

The definitions of the term "innovations" found in different sources may be grouped together according to the extent of the authors' approach to innovations. A significant group of authors perceive innovation as related to new technologies and to their employment in the creation and realization of new products and processes (Rajapathirana and Hui 2018; Terziovski 2007; Piktumiene and Kurtinaitiene 2010; Varkulevicius and Naudzius 2005; Alisauskas et al. 2007). In contrast, other authors propose a very broad understanding of innovations. They treat any new idea or changes and their successful realization as innovation without putting emphasis on the role of technologies in the innovation processes (Demirbas, Hussain, and Matlay 2011; Melnikas, Jakubavičius, and Strazdas 2000; Jakubavicius et al. 2003; O'Sullivan and Dooley 2009; Saatcioglu and Özmen 2010; Vijeikis 2011). Table 6.1 contains previously discussed concepts of innovations, proposed by various authors.

To sum up the ideas proposed by the foreign and Lithuanian authors, it is clear that innovation can be defined in different ways. Most of the scholars consider innovation to be an idea, an activity, a process, or a certain material object which is new to the companies, organizations, or societies which implement and/or use it.

In the modern, globalized business environment, it is becoming harder to create innovations "behind closed doors." Consequently, companies of different industrial branches have begun to open up their innovation processes and have obtained external resources and capacities (Wynarczyk 2013). Due to the development of information and communications technology and the growth of the internet, even small and newly found companies have managed to enter the global business arena (Onetti, Zucchella, Jones, and McDougall-Covin 2012). In such an environment, innovatory small and medium-size companies (SMC) play a crucial role between the local and the global economy, as well as between the dynamism and the competitiveness (Lecerf 2012).

Various authors classify innovations into different categories. For instance, Ohme (2002) sorts innovations into five categories: product, process, technological, innovatory, and profitable. According to the scholar, all these categories involve creativity, success, and inspiration.

On the other hand, Bigliardi and Dormio (2009) distinguish four types of innovation:

- product, which is a newly established service, idea or good;
- process, which is the implementation of new infrastructure and the application of new technologies, often leading towards the creation of new products;
- organizational, which encompasses the alterations occurring in the policy of marketing, administration, management, purchasing, and selling;
- marketing, which involves the creation of new markets or entering into already existing ones.

**TABLE 6.1**
**Concepts of Innovations**

| Definition of innovation | Source |
|---|---|
| Modern innovations are primarily based on the employment of new open technologies and open resources of high quality. Their basis consists of various new knowledge and information systems. | (Rajapathirana and Hui 2018) |
| An idea, practice or object which is individually understood as new. | (Demirbas et al. 2011) |
| In a broad sense, innovations are a creative activity, which transforms the scientific knowledge into new or improved products that are beneficial to the business and society. | (Vijeikis 2011) |
| The fulfilment of initiated actions which anticipate the required resources in order to achieve the desirable welfare. The application of innovations creates a competitive advantage which includes new technologies and methods. In a broad sense, innovations are the entirety of technical and social processes, the application of which results in the economic and social benefit. | (Saatcioglu and Özmen 2010) |
| Innovations are regarded as a process, during which new technologies, ideas, and methods are improved, together with their adaptation to the market. | (Piktumiene and Kurtinaitiene 2010) |
| Innovation is a large or small, radical or moderate execution of the changes in products or processes. It results in the emergence of something new which has additional value to the consumers and helps to increase the organizational knowledge. | (O'Sullivan and Dooley 2009) |
| Innovations involve the generation of new ideas and their fulfilment in the form of new products, processes, or services which have an effect on both the growth of the national economy and the rise in employment, and the increased profit to the company which implements the innovations. | (Jakubavicius et al. 2003) |
| Innovation is the application of resources by creating the value for the customer and the company and by improving and commercializing new and existing products, processes, and services. | (Terziovski 2007) |
| Innovation is a successful commercial application of new technologies, ideas, and methods by presenting new products to the market or by improving the existing products and processes. | (Alisauskas et al. 2007) |
| Innovation is the technological implementation or (and) development of new products and processes and it almost always involves obligatory technological modifications in the companies. | (Varkulevicius and Naudzius 2005) |
| Innovation is the renewal and the expansion of the products, services, and the interconnected markets. It is the implementation of new methods of production, provision, and distribution. It also includes alterations to the management, work organization, work conditions, and labor skills. | (European Commission 2004) |
| Innovation is a functional and advanced novelty which, in essence, is oriented towards the shift from old to new. | (Melnikas et al. 2000) |
| Innovation is a process during which a possibility is converted into new ideas and during which those ideas are widely applied in practice. | (Tidd, Bessant, and Pavitt 1997) |

*Source:* created by authors.

Rodriquez (2014) proposes a broader classification of innovations. He distinguishes innovations implemented in the fields of science, services, or production, and differentiates the latter ones into technological and non-technological. Nevertheless, the classification of innovations depends on the context in which it is used and on the broadness of the approach to the topic. The Oslo Manual (OECD/Eurostat 1997) presents a detailed classification of innovations (Figure 6.1).

Thus, innovations may be categorized based on various criteria: according to the chosen solutions to improve products or services; according to the decision to develop an existing or future business; according to the purpose of innovations; and according to the sector in which they are applied.

According to the first criterion, three types of innovation of a new product or service may be distinguished. They are as follows: continuous, dynamically continuous, and discontinuous innovations.

Continuous innovations include minor changes which are employed constantly. They help companies to distribute the investments and to stay competitive for a long period of time. The customers may not notice the alterations right away. However, the continuous application of innovations allows the company to maintain the existing demand since, during a long period of time, the needs of customers change, and the

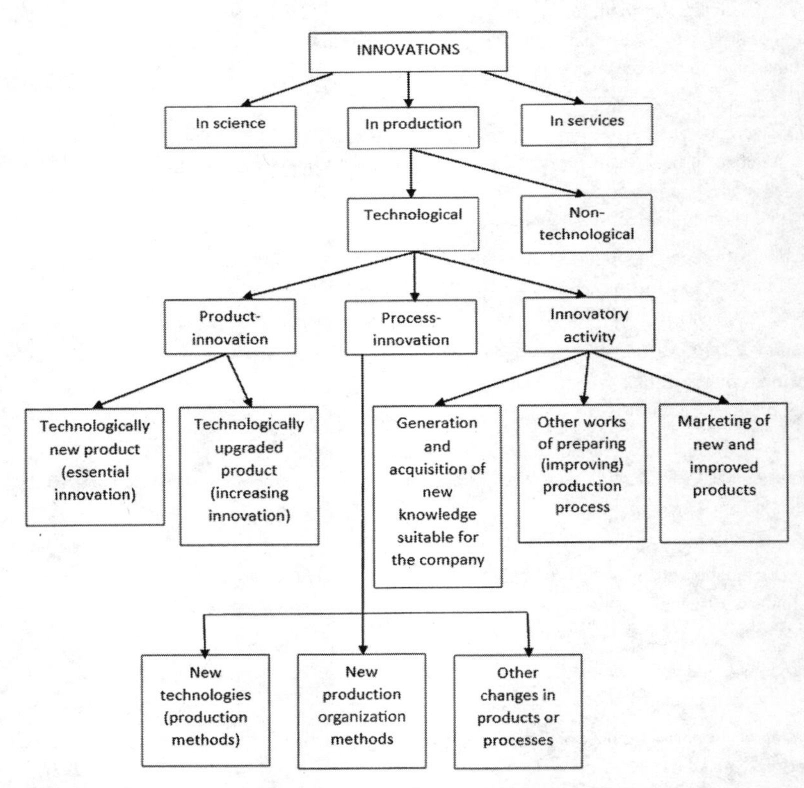


**FIGURE 6.1** Innovation classification according to OECD methodology. (Source: created by authors based on the Oslo Manual (1997)).

market competition increases. Such implementation of innovations not only economizes the investments due to the constant innovation application, but also helps to stay in the competitive market. This type of innovation covers minor changes to the product, which are related to its commercial design, and novelties, which are new in the internal environment of the company but familiar to the customers. According to Gammoh and Voss (2011), such products create new features which provide additional benefit to the existing value.

Dynamically continuous innovations are the invention of a new product, service, or modification (Saaksjarvi 2003). The creation of modifications of service or products does not require cardinal technical alterations since those modifications involve the development of already existing products. Contrary to the continuous ones, dynamically continuous innovations change the existing practice of use; therefore, it is necessary to anticipate the possible future needs, the fluctuation in demand, and the regularities of the competitive environment. According to Gammoh and Voss (2011), dynamically continuous innovations involve the new technologies and their implementation, the result of which is observed in the new infrastructure of the market. Innovations of this type are like substitutes of the existing products, which may alter the consumers' behavior.

Discontinuous innovations are characterized by large technical changes which radically separate the innovations from other products in the market. In most cases, they include the creation of new products or services, radical alterations to the existing production, generation of new ideas in the company's activities, and new inventions which influence the company's work in a modernized manner. Marketing specialists must prepare potential customers so that they properly react to the novelty. Only afterwards they should focus on selling the products since the new product has functions the previous merchandise did not have.

When classifying innovations according to their purpose, it is important to stress that innovation must change already established things by implementing modifications to the products or services and by inventing and proposing new models to the markets. All of this is understood as a development innovation and encompasses strategic processes for a long period of time. Hence, according to their purpose, innovations may be classified into eight types (Fifield 2008):

- Disruptive innovation is implemented when new technologies destroy or alter established things in the market.
- Application innovation is the employment of technologies in the new markets.
- Development innovation is understood as a strategic process, productivity and efficiency.
- Experimental innovation is a minor modification which enhances the customer experience.
- Business model innovation occurs when a new model, enhancing the benefit for the customer, is proposed.
- Structural innovation is the employment of policy changes in order to frequently reorganize industry processes.

- Marketing innovation is the improvement of services and of the relationship with the consumers.
- Product innovation occurs when a product or a service is proposed to the new markets.

According to their purpose, innovations can also be classified into intelligent and creative innovations. Intelligent innovations stimulate the emergence of the novelty, whereas creative innovations help to present it to the customers.

In the third group, innovations are categorized by the sectors: energetics, tourism, etc.

Having analyzed various scientific literature on the topic of innovations, it may be concluded that no universal definition, concept, and unified classification of innovations exist.

In summary, business innovations primarily are the commercial application of novelties by offering new products or services to the market or by improving the existing ones (Jakubavicius et al. 2003). Modern innovations are based on open new technologies and open resources.

It must be stressed that the implementation of innovations and the management of this process require preparation and detailed planning. First and foremost, innovation management needs a strategy and it encompasses such aspects as the organizational mission, the search of unique opportunities, their integration into the organization's strategic plans, the determination of success indicators, and, finally, the revision of unique possibilities.

Innovations can be realized only when the company has the capability to innovate. Innovative capabilities are considered to be a valuable non-material resource, which determines the company's long-lasting competitive advantage and the success of the fulfilment of the entire innovation strategy. Thus, the ability to innovate allows the companies to quickly introduce new products and to apply new systems. Innovation implementation requires a lot of resources, material and non-material funds, and innovative capabilities in order to ensure success in a rapidly changing business environment. Therefore, not all companies have proper capabilities to innovate.

According to Adler and Shenbar (1990), innovation capability consists of these parts: (1) capability to create new products which meet the market requirements; (2) capability to apply the necessary and suitable technological processes to produce new goods; (3) capability to create and adapt new products and production technologies to fulfil the future needs; (4) ability to respond to the unexpected actions of the competitors related to the new technologies and new products. A properly chosen strategy of innovation development is also one of the most important factors in successfully realizing the company's innovations.

## 6.2 INNOVATION STRATEGIES AND DEVELOPMENT FACTORS

Innovation strategies are a rather recent object of scientific research. In the context of a transitional economy, innovative strategies encompass two theoretical approaches:

the systemic level theories (Lundvall and Johnson 1994; Cooke, Heidenreich, and Braczyk 2004) and the organizational level innovation theories (Jaruzelski and Dehoff 2010).

The first movement of systemic level theories stresses the most important variables of innovative activities and focuses on the parameters of different innovative levels. The basis for these theories is the assumption that innovative activity of organizations is primarily determined by the national, regional, and (or) sectoral innovation system and that the activities related to innovations, as well as their typology and intensity, have a direct connection with the characteristics of the manifestation of the entrepreneurial organization dimensions in the country's institutional, juridical, organizational, and cultural infrastructure (Jucevicius 2007). This fundamental presumption is grounded on three very important theoretical concepts: *national innovation system* (Lundvall and Johnson 1994), *regional innovation system* (Cooke et al. 2004), and *variety of capitalist systems* (Hall and Soskice 2001).

Another movement of innovative strategies theories grounds itself on the organizational and entrepreneurial dimensions of the innovative process. The representatives of this movement make an assumption that innovations are developed not due to the surrounding factors (institutions), but due to the actions and decisions taken by entrepreneurs and organizations. In a sense, business organizations are not only affected by the external factors but may even shape some of them (Jucevicius 2007).

Hence, in scientific research, the approach to the relation between innovative strategies of organizations, their external environment, and the most important factors of shaping innovation strategies differs. The representatives of institutional economics or economics sociologists introduce organizational choices as an integral part of the external environment. The scientists of management and business fields study organizational innovation strategies as potentially independent from external factors and give priority to internal organizational determinants (Jucevicius 2007).

Combining the insights of the two theoretical movements, it may be stated that the factors of organizational innovation strategies include all interrelated determinants: the external environment, the internal characteristics of an organization, and the motives for creating innovation strategies.

Scientists do not have an agreement on the explanation of innovation strategies, their classification, and the definition of the determinants which influence the success of innovative strategies. In their study, Salaman, Storey, and Billsberry (2005) claim that innovation strategy is the creation of new products or business or service models in order to bring benefit to the company, the customers, and the partners. According to Zonooz, Farzam, Satarifar, and Bakhshi (2011), innovation strategy is a plan made to identify the strategic role of the new product which it will play in realizing the organization's goals.

Andersson, Johansson, Karlsson, and Lööf (2012) explain the innovation strategy in a very similar way. They describe it as a plan which is used by the company in order to promote progress in the areas of new technologies or services, usually by investing in scientific research and development. Therefore, in a broad sense, innovation strategy can be defined as a plan to implement novelties in a certain area, relevant to the company.

Scientific literature offers different classifications for innovative strategies. One of the older classifications for innovation strategies is proposed by Neely and Hii (1998). They divided them into three groups:

1. Offensive innovation strategy is characteristic of companies which are first to create a new product and present it to the market.
2. Defensive innovation strategy is applied by companies which aim to maintain their positions in the market. Organizations which adopt this strategy seek to preserve the achieved level and to create a niche in relatively stable product markets.
3. Imitative innovation strategy is attributed to companies which attempt to modify an already existing product.

Dogson, Gann, and Salter (2008) propose to classify innovative strategies into four groups: proactive, active, reactive, and passive. This categorization resembles the classic model and is grounded on how much the companies are prepared to create and implement novelties themselves or to attempt to apply already existing ones by copying more advanced companies. Companies rarely follow an ideal innovation strategy. However, the types of ideal innovation strategies might become a starting point in developing a company's own strategy.

Companies which apply a proactive innovation strategy have a stable direction of scientific research, hold the advantage of initiator, and are the leaders of the technology market. These companies make use of knowledge from various sources and take big risks. Businesses like *Apple* employ a proactive innovation strategy. Types of technological innovations used in a proactive innovation strategy are radical and sporadic. Radical innovations are understood as discoveries on how to change the nature of a product or service. Sporadic innovations promote to develop a product or service through the alterations in technological processes (Dogson et al. 2008).

An active innovation strategy is based on the protection of the existing technologies and markets. Companies which employ this strategy apply sporadic innovations. These organizations also possess large knowledge resources and take average risks. They tend to insure themselves. Companies like *Microsoft*, *Dell*, and *British Airways* belong to this category (Stoneman and Battisti 2010).

A reactive innovation strategy is applied by companies which are followers. They take their time and search for the smallest risks possible. These companies employ sporadic inventions; this group includes airline companies, for instance, *Ryanair*, which has successfully copied the service model of *Southwest Airlines*.

Businesses which apply a passive innovation strategy wait until their client's demand changes in the products or services.

Taking into consideration classifications of innovation strategies proposed by various authors, Stankevice (2014) divided the categories of innovation strategies into six groups according to the innovation types, innovation objectives, and innovation nature, and defined their competitive degree and their manifestation in separate industrial sectors:

1. semi-open, knowledge-receptive leadership innovation strategy;
2. expansive, marketing-based leadership innovation strategy;
3. product marketing and scope-based follower innovation strategy;
4. process and money-oriented incremental innovation strategy;
5. transformative innovation strategy;
6. corresponding, service-oriented innovation strategy.

A semi-open, knowledge-receptive leadership strategy is a very competitive innovation strategy, based on the improvement of communication and information distribution by decreasing the risk and the degree of technological, process, market, and other uncertainty. This strategy involves the creation of innovative products or services by conducting scientific research and experimental development in the organization itself and by employing the external sponsorship and the knowledge of universities and research institutions. This type of innovation strategy is applied in computer, electronic, and optical equipment, and in car and other industries.

A semi-open, knowledge-receptive leadership strategy could be ascribed to offensive (Neely and Hii 1998) and proactive (Dogson et al. 2008) types of innovation strategy formation. However, here the emphasis is put on the openness which is not examined in detail by other authors. Nevertheless, nowadays the openness degree of the innovation strategy has emerged as one of the most important classification criteria (Rajapathirana and Hui 2018).

An expansive, marketing-based leadership innovation strategy may be assigned to offensive (Neely and Hii 1998) and active (Dogson et al. 2008) types of innovation strategy formation. Its primary goal is to occupy new markets by increasing the capacity of goods and service production. It is a rather competitive strategy and it involves the development of new pricing and work methods, as well as external relations, by proposing new original solutions to the market. This strategy is most common in the fields of information and communications technology, programming, telecommunication, and in the wholesale and retail trade. Its distinguishing characteristic is innovative marketing.

A product marketing and scope-based follower innovation strategy can be ascribed to imitative (Neely and Hii 1998) and reactive (Dogson et al. 2008) types of innovation strategy formation. Its goals are to update outdated products, to improve their quality and diversity, and to attract new customers. This strategy includes the development of the pricing of new products, as well as the development of external relations and the collaboration with other organizations. Organizational innovations are mostly new to the companies themselves, but not to the market. The discussed strategy is most prevalent in the wholesale and retail trade, in the food and drink industry, and in the field of information and communications services. It is distinguished as a strategy of low-competitiveness.

According to Stankevice (2014), a process and money-oriented incremental innovation strategy indicates the development of new products by improving logistics systems. It may be assigned to defensive (Neely and Hii 1998) and passive types of innovation strategy formation since companies produce goods which were created in other organizations. These decisions are innovatory only to the organizations themselves. This strategy is mainly observed in the areas of finance and insurance, the sectors of chemistry and pharmacy, and in the industries which produce rubber, plastic, and other oil products. It is a moderately competitive innovation strategy.

A transformative innovation strategy is directed towards the reduction of costs, the increase of the market shares and the organizational flexibility, and the assurance of production scope economy by improving information distribution, product quality, and seizing new markets. It may be ascribed to imitative and reactive types of innovation strategy formation. When applying this strategy, organizations most frequently implement new management systems and coordinate the strategy with other innovation strategies. This type of innovation strategy is observed in the industries of wood, straw, and other related products, as well as in the fields of computer, electronic and optical, car and transport equipment production. It is a moderately competitive innovation strategy.

A corresponding, service-oriented innovation strategy may be grouped under defensive (Neely and Hii 1998) and passive (Dogson et al. 2008) types of innovation strategy formation. It is based on the provision of new or significantly modified services and its main task is to reduce the time to reply to customers and suppliers. This strategy does not include essential alterations in product design or packaging. It is applied in the fields of finance and insurance services and in the sector of information and communications technology. It is distinguished as a lowly competitive strategy. Nonetheless, it is considered a better alternative in comparison to product marketing and scope-based follower innovation strategies (Stankevice 2014).

Out of the discussed strategies, the semi-open, knowledge-receptive leadership innovation strategy may be regarded as one of the most competitive innovation strategies. It is an offensive, proactive innovation strategy; however, not every organization can select to implement it. It depends on the circumstances in the country, the sector, and the industry, and on the external and internal factors, as well as the company's situation in a particular market, the size and novelty of the market, and other reasons.

It must be noted that, currently, the openness (of various degrees) is one of the most significant aspects in innovation strategy formation. The degree of strategy openness is closely related to the company's business model, organizational system, culture, and other important aspects. Open or semi-open innovation strategies, chosen by the modern leader companies, make it possible to employ open new technologies and high-quality knowledge resources even for SMC which do not have large finance and human resources.

In addition to types of innovation strategy formation, authors concentrate on the model of innovation strategy development. These models can be classified into the following main groups (Alisauskas et al. 2007): (a) linear models, and (b) cyclic, systematic models.

A linear model of innovation strategy development presents a traditional vision of knowledge generation. It clearly distinguishes between science and technological expertise and private and public scientific research, where scientific knowledge is considered to be a public product and knowledge equals information. This model may be described as consecutive emergence of new processes and products (Alisauskas et al. 2007; Bandzeviciene 2011).

Figure 6.2 presents the linear model of innovation strategy development.

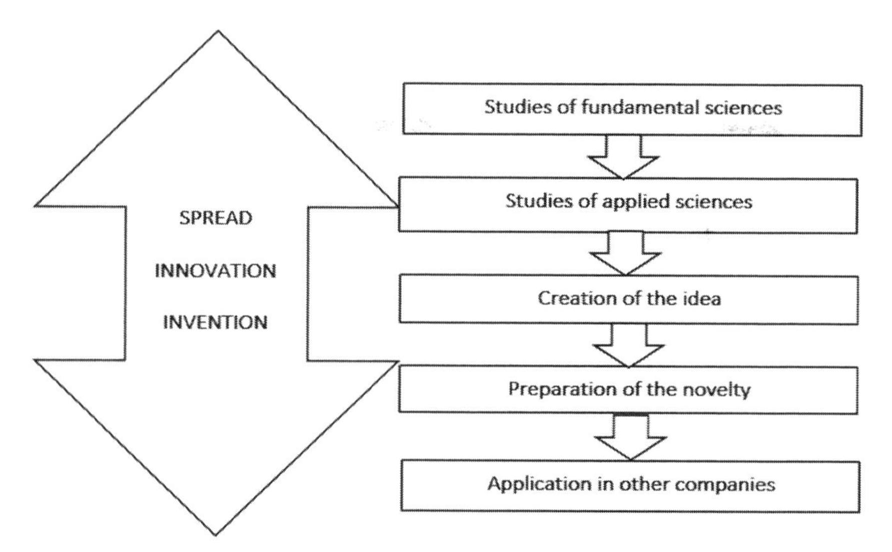


**FIGURE 6.2** A linear model of innovation strategy development. (Source: created by authors based on Alisauskas et al. (2007) and Bandzeviciene (2011)).

The development of innovation strategies may be fulfilled according to the cyclic model, which significantly differs from the linear one. The essence of the cyclic model is the knowledge which comprises the positive environment for the development based on the research, the analysis of potential markets, and the designing of the product and the service (Zizlavsky 2013; Sapiegiene, Jukneviciene, and Stoskus 2009). Figure 6.3 contains the cyclic model of innovation development.

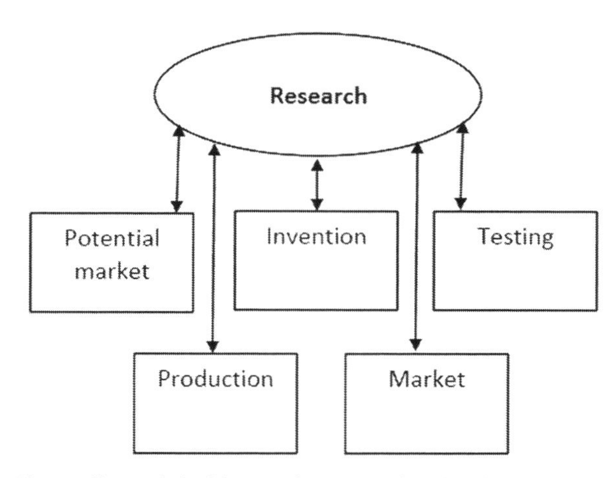


**FIGURE 6.3** The cyclic model of innovation strategies development. (Source: created by authors based on Stoskus et al. (2005) and Zizlavsky (2013)).

The possible models of innovation strategies development are summarized in Table 6.2.

It must be stressed that the market is the criterion for the success of innovation creation. Society as a consumer of services and products can control the process of innovation creation. Moreover, the cyclic model of innovation strategy development is grounded on the continuous dialogue with society (Sapiegiene et al. 2009).

Other scholars distinguish the following models of innovation strategy development (Terziovski 2007):

- Technology push model. It is a first-generation innovation epoch and it was the basis for the industrial revolution. Innovations were developing with

**TABLE 6.2**
**Model of Innovation Strategies Development**

| Model | Description |
|---|---|
| A linear model of technological networks | By applying this model, companies look for various innovation "agents" through collaboration networks and information exchanges. Here an important role is played by the information sources outside the organization: the clients, suppliers, consultants, experts, science institutes, universities, state institutions, and others. |
| A linear model of social networks | In this model, innovations are influenced by scientific research and by the informal interaction between organizations and other participants (theory of technological network).<br>The main novelty of this model is the crucial impact the knowledge has on the innovations. The growth of knowledge importance is related to the information distribution and the development of communications technology which makes the knowledge easily accessible. |
| Cyclic model of technological stimulus or engineering | In this model, innovations arise from the science, i.e., innovation sources are related to scientific research, scholars' activity, and technological progress.<br>Companies discover innovation opportunities by using the results of scientific research.<br>Studies of fundamental and applied sciences are the primary sources of the emergence of new and advanced products and processes in the companies. |
| Cyclic model of market inclination | When applying this model, innovations are created by taking notice of and reacting to the market needs since they generate new ideas and solutions. The main source of knowledge on how to improve a product or processes is scientific research. Furthermore, organizational factors have an important role in the innovation creation because innovation implementation requires not only technically available solutions but also organizational competence. |
| Transactional cyclic model | In this model, innovations originate from informal, spontaneous interaction between the participants of the market. It creates more chances to establish relations between science and business, research and market, marketing and selling, and to develop innovations by employing ideas circulating in the surrounding environment and information received through existent connections between companies, customers, suppliers, and others. |

*Source:* created by authors based on Alisauskas et al. (2007) and Zizlavsky (2013).

new, technologically advanced products and new industries. Such advanced products were pushed to the market.

- Need pull model. It is a second-generation innovation epoch and it was directed towards customer needs. The growth of customer needs also determines the rise in production technologies. In this epoch, the market is the main generator of new ideas.
- Coupling model. It is the third innovation epoch and it combines the first two models of innovation strategy development. The market was in demand of new ideas, and the need for innovative products and services was increasing; however, at the same time production technology was in development, new fields of industry were emerging together with advanced technologies.
- Integrated model. In the fourth-generation integrated innovation model, the emphasis is put on the close connection of marketing and scientific research and technological development (SRTD) with the suppliers and customers.
- Systems integration and networking model. This fifth-generation model is based on the strategic partnership with the suppliers and customers. Here expert systems, marketing, and research integration are implemented.

Some of the discussed models are suitable for certain industrial branches depending on the context. For instance, the model of technology push is more fitting for pharmacy, whereas the model of need pull is more frequently applied in the industries of wide-usage products (Nazali, Noor, and Pitt 2009).

Lately, a few main models have been in development: the method of comparison of national innovation systems, the first and the second methods of scientific knowledge integration, and the "triple spiral" model. The latter is considered to be the most advanced of the mentioned ones. It is based on the collaborations of science, business, and state institutions. The principal idea of the model can be described as follows: in the society of knowledge economy, the boundaries between the public and the private sectors, science institutions and business are disappearing, giving opportunity for the formation of a new system, the elements of which intertwine and strengthen one another (Chlivickas, Petrauskaite, and Ambrusevic 2009).

In regard to the models of innovation strategy development, it must be stressed that the development of innovation strategies is a complex and dynamic process since the connections between science, innovations, and economic development are interrelated and recurring. Therefore, they are best explained by referring to the non-linear models of innovation strategy development.

When selecting a particular type of innovation strategy formation and its development model, an important part is played by the external environment and the company's internal circumstances, the significant elements of which may encourage or interfere with the innovation processes. Together with promoting and hindering factors, it is necessary to choose a type of innovation strategy formation which would allow the company to successfully overcome various barriers and to take advantage of favorable factors or motives.

A relatively big group of scholars has also analyzed the motives of innovation strategy implementation (promoting factors) and barriers (hindering factors). Two main directions can be distinguished: (1) the classification of innovation development factors into external and internal; (2) the grouping of innovation development factors according to other features.

Most scholars divide the strategy barriers and motives into internal and external. Vijeikiene and Vijeikis (2000) investigate the following main (internal) motives of innovation implementation: organizational structure; knowledge; resources; management; quality need; personnel; company's culture; and desire to take advantage of the privileges of innovation promoting policy. Whereas the external motives of innovation implementation, as pointed out by these authors, are as follows: political decisions; technological progress; changes in the society; environmental and social requirements; partners and suppliers; and market changes.

There exist not only the motives of innovation implementation but also the barriers. They can be external and internal. Market disturbances and the competitors' response to the innovations are distinguished as the external barriers of innovation implementation. Fierce competitive rivalry, deficiency in suppliers, the inconsistency of raw material prices, and difficulties in finding collaboration partners also hinder the innovation implementation in organizations (Sapiegiene et al. 2009). According to Skarzynski and Gibson (2008), internal barriers of innovation implementation are the lack of information; fear of risks; shortage of new ideas; resistance to novelties; process management; the leader's poor competence and knowledge; and lack of financial and human resources. External barriers of innovation implementation are market disturbances; the competitors' response to innovations; difficulties in finding collaboration partners; the lack of political stability; and absence of state support.

According to Saatcioglu and Özmen (2010), internal factors and barriers of innovation strategies are related to certain resources (finances, technologies, time), while external barriers are linked with the provision, requirements, needs, and suitable environment, which consists of raw material, technological knowledge, dangers of innovations, laws of local and foreign markets, restrictions, and political issues. Hence, human resources, government policy, and finances are primarily related to the barriers of innovation implementation. As stated by Madrid-Guijarro, Garcia-Perez-de-Lema, and Van Auken (2009), the most dangerous are the internal barriers, such as the lack of financial resources, which has a direct negative impact on human resources and creates various risks. Muller-Camen, Croucher, and Leigh (2008) also focuses on the internal barriers of human resources related to abilities and motivation. In his opinion, barriers of human resources related to competence, communication, and resistance to innovations arise from the lack of knowledge and motivation. A company's external environment includes government policy and various global factors. Companies which wish to gain or maintain a competitive advantage must accept innovations and implement innovative processes. However, the lack of information may become a large barrier which prevents innovation adaptation in the market (Madrid-Guijarro et al. 2009). This type of barrier is closely related to the national restrictions, cultural, market entrance, and partner search barriers.

Summarizing the first scientist group's research on innovation promoting and hindering aspects, the following main internal and external factors determining the choice of innovation strategy type can be distinguished: finances; human resources, government policy; access to technological providers; information accessibility; and market size and competition.

The second direction of the classification of the factors of innovation implementation identifies the subsequent innovation development factors: relative advantage; compatibility; complexity; the possibility to test; observation; and risk (Kale and Arditi 2010; Christou, Avdimiotis, Kassianidis, and Sigala 2004). Here the focus of attention is on the analysis of innovation development barriers, although some development factors might be promoting. Relative advantage is a barrier which may be relationally measured and evaluated. It is estimated on the basis of influence, loyalty, and satisfaction (Christou et al. 2004). Relative advantage can be determined with economical, technical, and social factors. Compatibility is the second barrier, preventing the adaptation of innovations. Some authors (Kale and Arditi 2010; Christou et al. 2004) state that innovations' compatibility with the existent values, convictions, traditions, experience, and needs in the social environment raise a lot of problems in the implementation of innovation strategies. Complexity is the barrier related to the features of the product itself. Easily employed novelties will be accepted more rapidly than those which require new skills, knowledge, or appropriate understanding (Tapaninen 2008). The observation barrier is linked with the novelty advantages and their noticeable exclusive features which increase the demand (Christou et al. 2004). The possibility to test raises the dangers of the perceived risks and lessens the probability to the selling company that the novelty will be accepted (Tapaninen 2008). Such strategy is applied by the companies which have confidence in their innovation and its future in the market. All the above discussed barriers may be considered as risk barriers since they signify a certain risk to the customer.

To sum up, both the first and the second directions of classification of innovation strategy development can be applied by selecting a particular type of innovation strategy formation. However, the division into internal and external factors is more universal since, in the second direction, factors of innovation development are analyzed as barriers. Moreover, they depend on the relations between the external and internal factors. Taking into consideration the complexity of the investigation of innovation strategies, classification into the external and internal factors makes it easier to structure the research and to present a clearer analysis model.

Having analyzed the scientific literature, it might be concluded that the successful implementation of innovation strategies is dependent on the company's internal and external environment factors which are divided into hindering (barriers) and promoting (motives). The main barriers of innovation implementation are the lack of financial and human resources and the instability of the economic and political environment. The most important motives of innovation implementation are continuously changing demand and the increase in competitive advantage.

## 6.3 SUSTAINABLE INNOVATIONS AND THEIR PROMOTION

In their study, Carrillo-Hermosilla, Del Río, and Könnölä (2010) define sustainable innovation on the basis of a concise concept of eco-innovation. It is an innovation which improves the performance of sustainability. According to the scholars, such performance includes ecological, economic, and social criteria. It is observed that very often scholars place sustainable innovation close to eco-innovation and stress its characteristics. Ecological and economic dimensions of sustainable innovation are directly related to the circular economy and its activities.

Other authors (Boons and Ludeke-Freund 2013) do not limit their definitions of sustainable innovation to the concept of eco-innovations since sustainable innovations encompass social objectives and are clearly directed towards the holistic and long-lasting process of sustainable development, including short- and long-term objectives of sustainability. In the context of sustainable development, the presented concept of sustainable innovation is observed in the long and short-term goals of sustainable development, shared with some goals of the circular economy, such as zero waste production, recycling, re-use, preservation of natural resources, and others.

According to other authors (Charter, Gray, Clark, and Woolman 2008; Charter and Clark 2007), sustainable innovation is a process during which aspects of sustainability (environmental, social, and financial) are integrated into the company's/organization's system by moving from idea generation to scientific research and development, the final step being commercialization. It is applied to products, services, and technologies, as well as to new business and organizational models. The notion of sustainable innovation emphasizes some of the goals of sustainable development; however, these two conceptions diverge in one essential difference—social dimension.

The social dimension is very important to sustainable innovations. The insights of Pusinaite (2015) confirm it. According to her, the types of sustainable innovations are direct eco-innovations, encompassing eco-innovations which have a positive private economic result and an indirect social effect, and eco-innovations which have a positive private and external economic result and a clear social aspect. The scholar (Pusinaite 2015) distinguishes direct social innovations as a type of sustainable innovation. They include social innovations which have a private economic result and an indirect environmental effect and social innovations which have a positive private and external economic result and a clear environmental aspect.

In their study, Carrillo-Hermosilla et al. (2010) define sustainable innovation as eco-innovation by taking into consideration many different concepts. However, very often eco-innovation is used as a synonym for sustainable innovation. Related definitions, for instance, cleaner technologies, are used as partially coinciding with the ecological performance and characteristics. This situation arose as a result of the fact that scientists of various fields chose to analyze this topic: evolutionary economics, science and technologies, innovation economics, economic sociology, and history. Most of these fields focus on innovations which have an ecological effect on a product or service (Boons and Ludeke-Freund 2013), apparent in the context of the circular economy.

In the examination of the contribution of sustainable innovation studies to the systematic analysis of literature, four factors directly related to environmentally sustainable product innovations are distinguished: market, law, and regulation knowledge; interfunctional collaboration; innovation-oriented learning; and investments into scientific research and development (De Medeiros, Ribeiro, and Cortimiglia 2014). All these factors have a direct connection to the realization of a circular economy.

Based on insights of Pusinaite (2015), sustainable innovation can be considered as an organizational activity which creates economic benefit and has a positive effect on the environment and society. In such a context, it is possible to detect relations with the circular economy since it also has a double objective—economical profit and the lesser effect on the environment. Therefore, an assumption can be made that the circular economy may be realized through the implementation of sustainable or eco-innovations.

The importance of sustainable innovations in creating a circular economy is also emphasized in the priorities of the European Commission's Horizon 2020. One of the main problems discussed in the program is waste which is the source for recycling, reuse, and raw material restoration. The program indicates the direction towards the circular economy (described as restoration economy) through industrial symbiosis as a priority. Furthermore, a systematic approach to reduction, recycling, and reuse of food waste and to building remaking into raw material must be created. Another priority is identified as a direction towards near-zero waste production on the European and global levels. The promotion of eco-innovative waste disposal and waste prevention is also accentuated as a part of sustainable city development (Horizon 2020). Thus, it can be observed that the movement towards a circular economy in Europe is promoted through eco-innovations or sustainable innovations. Therefore, in Horizon 2020, eco-innovations are considered as the main engine for creating economic growth and new employment in Europe and the world.

The factors of sustainable innovation promotion can be of several types. They can be external (Yoon and Tello 2009) and internal (De Medeiros et al. 2014), economic and non-economic (Pusinaite and Pucetaite 2015).

Yoon and Tello (2009) distinguish the following main factors promoting innovations:

- Customer demand. Customers begin buying ecological, environmentally friendly services and products. In such a way, changes in business practice are created.
- Social responsibility initiatives. Companies begin demonstrating their commitment to social responsibility. By doing it, the company's competitiveness increases and exclusiveness in the whole industry is created.
- Government intervention. It ensures the conformity of legal acts, promotes a positive effect on the environment, and creates political economical means to promote technological progress.
- Social movement. It enhances the social awareness to the environment and social justice and, thus, influences customer demand and redistributes political forces between the state government and business practice.

- Technological progress. By developing efficient technologies, wasting of raw material and pollution are diminished and new business opportunities and initiatives are supported. Technological progress also increases the company's competitiveness and creates exclusiveness in the whole industry.

All these factors are external; therefore, companies cannot control them. In this regard, the government and its policy have more influence—for instance, by educating and informing customers and, thus, increasing customer demand and creating political means for promoting technological progress.

In their study, De Medeiros et al. (2014) distinguish four variables as a means to promote sustainable innovations:

- Market, laws, and knowledge of the regulation system. Sustainable innovation promotion and development depend on the customer readiness and capability to accept such innovations, on the ecological legal acts, government initiatives, and educational campaigns, spreading sustainable development culture in the society.
- Intersectoral collaboration. It indicates that synergy between different sectors must occur not only inside but also between the interested parties participating in the creation of ecological products and in the preparation of their presentation processes.
- Innovation-oriented learning. This type of learning is dependent on cultural barriers. These barriers may become an obstacle in taking advantage of market capabilities related to sustainable innovations promotion and creation. Therefore, innovation-oriented learning must be taken into consideration.
- R&D investments. Companies which wish to develop and exploit environmental sustainability must invest in scientific research, develop new technologies, upgrade their production systems, and improve processes of product development.

Pusinaite and Pucetaite (2015) classify factors of sustainable innovation promotion into economic and non-economic. Economic promotion factors include:

- Tax concessions to organizations which implement sustainable innovations. This promotion factor can be applied only when the new product is desired to become more competitive than the traditional ones.
- Partial sponsorship of sustainable innovation implementation. Scholars consider this promotion factor to be the most frequently mentioned economic means of sustainable innovation promotion.
- Funding for infrastructure projects. Capital invested in infrastructure must bring value. What is more, the infrastructure must be adjusted to be able to solve social and environmental problems and has to be supported financially continuously.

- Redistribution of funding for the government and the social business. This means of sustainable innovation promotion emphasizes the importance of evaluation of who can employ the obtained funds better and more efficiently—the government or the sector of private business.
- Support for local initiatives. Purposive funding to particular locations may be more relevant and efficient since, for instance, country-side communities know best the problems which need to be solved in their region.
- Formation of a larger demand and products created with sustainable innovations by employing public procurement. It is identified as a sustainable innovations promotion factor for social business.

The authors (Pusinaute and Pucetaite 2015) classify the following non-economic factors:

- Clear and consecutive government policy of sustainable innovations sponsorship. In regard to sustainable innovations, the government policy has to be directed towards the promotion of companies' management, the creation of new employment, and new industrial branches.
- Non-material support of government institutions. Small support is also suitable for the business: free of charge car parking in city areas, state officers' attention to the project, the decrease in bureaucratic procedures.
- Customer education. Customer education includes marking, various informative campaigns, and standardization.
- Strengthening of government regulations. Government regulation and supervision must be enhanced in the creation and development of sustainable (eco) innovations. To achieve optimal results, stricter legal instruments of regulation should be enacted.
- Implementation of new forms of collaborations with business and non-government organizations in the state sector.
- Long-lasting strategy in seeking to promote sustainable development of the society (e.g., promotion of family institution). The broadening of the concept of the social company could make it easier not only to solve various problems but also to support values, capable of preventing the issues.
- Changes in the education system. The implementation of sustainable innovations requires the preparation and development of specialists in this field.
- Promotion of alterations in cultural organizations. The cultural organization can become the power to initiate the changes in society.

Having analyzed the factors of sustainable innovations promotion, it is observed that they can be divided into internal (dependent on the company's activity and policy) or external (dependent not on the company's activity and policy, but on the government policy and regulation), as well as into economic (subsidies, privileges, funding, and support) and non-economic (formation of government policy, customer education, strengthening of regulation, long-lasting strategy, alterations in education system, and promotion of cultural changes).

## 6.4 GOALS AND INNOVATIONS OF SUSTAINABLE DEVELOPMENT

The provisions of sustainable development were most recently revised in 2015. Then, after three years of negotiations, the UN's agenda of sustainable development until 2030 was ratified. It encompasses 17 goals and 169 objectives of sustainable development which substituted the objectives of the Millennial development set 15 years ago. The latter objectives, having helped to lessen the poverty in the world with the mutual effort of many countries, were directed towards the developing countries and did not have a unified realization strategy.

The 17 goals of sustainable development agenda include the following objectives:

1. No to poverty: to end any kind of poverty in all parts of the world.
2. No to starvation: to stop hunger by ensuring food safety, by improving nutrition and by promoting sustainable agriculture.
3. Good health: to ensure the possibility for a healthy life and to encourage the welfare of all people of all age groups.
4. High-quality education: to guarantee diverse and impartial high-quality education and to promote learning accessible to everyone throughout their lives.
5. Gender equality: to achieve gender equality and to support women and young girls.
6. Clean water and sanitation: to ensure universal access to clean and properly regulated water.
7. Renewable and accessible energy: to guarantee universal access to affordable, reliable, sustainable, and modern energy sources.
8. Good employment and economy: to promote stable, versatile, and sustainable economic development, to ensure full and productive employment and a suitable job for everyone.
9. Innovations and good infrastructure: to create a flexible infrastructure, to promote universal and sustainable industrial development and to foster novelties.
10. Diminishing of inequality: to lower the inequality of separate countries.
11. Sustainable cities and communities: to make cities and settlements comfortable, safe, sustainable, and able to flexibly adapt to circumstances.
12. Responsible usage of resources: to ensure sustainable models of consumption and production.
13. Climate protection: to immediately take action to stop climate change and to eliminate its consequences.
14. Sustainable oceans: to protect the oceans, seas, and their resources and to use them responsibly by ensuring their stable evolution.
15. Sustainable use of land: to conserve and to recreate terrestrial ecosystems, to promote their sustainable usage, to responsibly manage forest resources, to fight against desertification, to stop the degradation of and recreate the soil, and to prevent the extinction of biological diversity.

16. Peace and justice: to support the development of peaceful and tolerant society, to provide all people with the possibility to protect their rights with legal means, and to establish multifunctional institutions operating in clean and efficient manner on all levels.
17. Sustainable development partnership: to renew global partnership in order to create sustainable progress and to enhance the means to realize it.

The goals of sustainable development are depicted in Figure 6.4.

The execution of the UN's agenda 2030 is recommendatory and its violation would be sanctioned. However, the governments are expected to take responsibility for the realization of all the objectives by creating or applying already existing national plans, programs, and strategies.

The sustainable development agenda 2030 acknowledges the different levels of each country's development and the different available resources and capabilities, and, thus, is a universal plan which leaves no person secluded despite their unique complicated circumstances (disability, isolation, poverty, and so on) or their capabilities to realize the goals.

The agenda does not require additional resources but encourages the redistribution of existing ones accordingly and, after evaluating the situation in every country, the search for supplementary sources from the set of 17 goals. Goals and objectives of sustainable development are integral and undivided. Although certain areas of priority may be distinguished, their realization cannot dominate over the implementation of others. For instance, economic growth (SDG8) may contribute to poverty reduction (SDG1) and to proper employment demand (SDG8). Other goals, such as responsible consumption and production (SDG12), city and community sustainability (SDG11), influence climate change (SDG13), and smaller isolation of less empowered people or poorer countries (SDG16, SDG17) cannot be ignored as well (SDG10). In most cases, high-quality education (SDG4), good health (SDG3), and gender equality (SDG5) are the basis for a

**FIGURE 6.4** Goals of sustainable development. (Source: created by authors based on United Nations (2018)).

steady progression of any other goal. Each country will have to present a report on the progress of realization of all the goals.

It is important for business, competitiveness, novelties, and market capabilities to move to a more sustainable practice. SDG must be a part of the business and financial community and to mobilize and involve them as change agents. It requires to unify limit narration and to emphasize possibility narration. Social and technological novelties are essential in preserving "safe and just space of activity" by opening new business opportunities and by promoting radical means of new products, services, and welfare provision.

The motivation of predominant advantage, the lessening of the risk of the global provision chain, the increase of efficiency, the changing norms and values of a company, and the stability have been included in the strategic plan of many dominating companies. The newest example of novelties in business participation and of risk management, related to sustainability challenges, is the integrated initiative of accountability, which is used to promote integral accountability of sustainability and finances as ratified in the Global Commission of Sustainable Development and Business program of 2020. Action or initiative "Risky business" evaluates the economic climate risk in the USA. More extensive efforts in regard to green competitiveness are incorporated in the inclusive agenda of green economy and in the recently announced new initiative of climate change, which is led by the former president of Mexico Felipe Calderón. These are the first steps. Nevertheless, in order to contribute to the business, enterprise and finance companies have to advance further and quicker. The governments have to determine appropriate encouraging means and good practice, corresponding with the sustainable development goals and objectives, and to promote sustainable systems and practice.

SDG may contribute to green competitiveness with new ideas and technologies on the macro-level and by promoting the new business practice. Technologies are included in the SDG realization (17.6–17.8 goals), mostly focusing on passing on the knowledge. This formulation of technological problems is first related to access to advanced technologies in developing countries and ignores the main issue of the necessity to promote the transformation of system novelties between economy and society. It is one of the sustainability conditions. The SDG agenda also contains recommendations for the task of providing reports of sustainable practice and companies (including the integrated ones). This goal (12.6 from the agenda) may attain distinct indicators. The indicators could be the number and the quality of set standards of sustainability and national legal acts which enforce the application of integral accountability and sustainable practice of acquisition. The business has the main role in the process of sustainable development and must be directly accountable. Efficiency goals of sustainable production and resources use, directly related to companies, not only will support the dominant leaders in creating sustainable business models but also will put pressure on other companies to change unsustainable business practices.

Innovations have always played a significant role in economic development. Eventually, the growth of one person's income is determined by the efficiency changes, closely related to technological progress and novelties. Even in a moderate and short period of time, a big benefit can be achieved by integrating modern technologies and new practice in developing countries. For instance, access to information and communications technology (ICT) improves life quality in most remote world regions since it allows people to interact, learn, and manage their business in an efficient way. Biotechnologies and precision agriculture, performed by such geographical places, drones, smart sensors, and cloud computing, may improve fertility and farmers' living conditions in regions which experience negative climate conditions. Renewable energy technologies demonstrate an impressive pace and provide more extensive access to electricity, which is an essential condition for efficient modernization and a means of implementation of many other sustainable development goals—for example, good health, gender equality, and high-quality education. Environment protection technologies help to fight against harmful pollution and to improve energy efficiency in production sectors. Without a doubt, technological novelties have a huge potential to bring benefit for companies, society, and the environment. However, the reasons for their potential are not the technological merits which are supposedly so vast that they force the implementation of technologies or the required scope, but the clear national policy, supported with the sufficient amount of national and foreign investments, and efficient mechanisms, which ensure the simplification and transfer of technologies. In any case, all novelties, including the ones which bring the largest benefit, encompass certain economic, social, or environmental compromises, which should be determined, evaluated, and solved.

It is a complicated task in which countries must be able to use the knowledge and institutional resources, which are still poor in many developing countries. Without these resources, it is difficult to encourage companies to innovate and to directly change the policy. Some countries have the drawbacks of small capabilities, little attention paid, the lack of funds, and actual state tax authority integration into the development strategy. It is a big challenge since latecomers are in an unfavorable position in competitions, which are based on cumulative knowledge and skills, long-term interactions, and complex material and non-material infrastructure. Therefore, developing countries should be supported by creating a consistent system which would join innovation efforts with trade, investments, competitiveness, and industry policies. In such systems, extensive participation of social and economic agents will be very important, considering proof that success in solving development problems by implementing novelties is related to the scope of the proposed solutions.

Hence, if achieved by 2030, SDG may not only build peace and share the welfare in a healthy planet but also open up possibilities to a US$12 million market. Exponential goals require transformative decisions which encompass not only progress. Companies must set their own sustainable objectives and firmly take action:

1. to accept a new mind-set;
2. to create and test new business models;
3. to create and implement technologies.

These "break-through novelties," as we name them, assist the companies in overcoming market disturbances and in seeking growth opportunities, while at the same time, helping to tackle global warming and to radically improve life.

# 7 Conclusions

Only sustainable development can ensure social welfare and happiness for residents of a country, and to solve the main economic, social, and environmental problems.

The implementation of sustainable development on the macro-level can be secured by realizing sustainable principles on the micro-level. Therefore, sustainable development of organizations, creation of a sustainable, socially responsible organization, and other transformations of companies into sustainable ones are some of the most important factors in seeking sustainable development of the country.

Sustainable, socially responsible, and high-culture organizations are distinguished by their stable financial indicators and sustainable development, they are concerned with the protection of the environment and nature resources, and they create high-quality employment for their employees. They also directly contribute to the solving of the community's social, environmental, and other problems, they increase the society's trust in business and the residents' solidarity and safety, while at the same time decreasing disappointment and depression and raising the population's psychological resistance and feeling of happiness.

Having assessed the huge benefits sustainable and socially responsible organizations bring to society, it is necessary to focus on the leaders of such organizations and to reveal and emphasize the importance of sustainable leadership in creating an organizational culture and in transforming it into a sustainable and socially responsible organization.

According to leadership theories, leadership is not only important to the organization or to the career of an employee, but it is also essential in all fields, countries, and communities. Leadership is understood as an exertion of influence in order to help the follower groups or single people to achieve significant goals.

Recently, changes in the management paradigm, where modern and traditional approaches are compared, are observed. A modern organization must focus on the importance of the contemporary approach, which is characteristic of leadership, viewing ahead, management of risks and knowledge, flat organization, empowering, self-control, teamwork, and learning oriented towards future needs.

To transform an organization into a sustainable one, it is recommended to employ sustainable leadership. Sustainable leadership is a sustainable and continuous process, grounded on the use of non-forced influence over the members of the collective with the power and the competence of the leader's personality, by attempting to direct or coordinate team members' actions and to form followers-leaders in order to achieve the objective, which is based on the organizational values, mission, and vision.

The presented concept of sustainable leadership is a holistic conception and encompasses the main dimensions of sustainability (environmental, social, economic, and institutional) and the main components of leadership process (interaction, mission, based on values and vision, result, and power of the personality), as well as most significant types of intelligence: rational, emotional, spiritual, and physical.

The proposed model of sustainable leadership implementation is suitable for organizations of public and private sectors and involves three main cycles of sustainable leadership process (commitment, collaboration, and change). Here, the most important role is given to the mission. The mission is based on the environmental, social, economic, and other values of an organization. It may be described as a formation of organizational culture by seeking to realize the most important vision—the creation of a sustainable organization.

Sustainable leadership, implemented according to the proposed model, can bring significant changes in organizational culture and allow the organization to fulfil its mission to transform into a sustainable organization whose main values include environment protection, preservation of natural resources, and social obligation to employees and society.

The analysis of good practice displayed by the sustainable and socially responsible companies operating in Lithuania has revealed that these companies have achieved very good results due to innovative technological and management solutions and all these companies are marked by the strong presence of a leader. The implementation of sustainable leadership principles is the factor which allowed the companies to reach high organizational culture and to ensure their sustainable development.

The creation and development of sustainable companies by employing principles of sustainable leadership would ensure the sustainable development of Lithuania, would guarantee the safety of the society, and would increase the residents' psychological resistance, which would improve the country's capabilities for sustainable development.

Sustainable innovations help companies to achieve the goals of sustainable development by applying the principles of leadership and sustainable management. The implementation of sustainable development on the macro-level can be achieved by the sustainable and socially responsible business which would employ sustainable innovations on the micro-level by applying principles of sustainable leadership.

# References

Adair, J. 1998. *Effective Leadership. How to Develop Leadership Skills?* 2nd Edition. London: Pan Books.
Adler, P. S. and A. Shenbar. 1990. Adapting your technological base: The organisational challenge. *Sloan Management Review, 25*: 25–37.
Adriaanse, A., S. Bringezu, A. Hamond, Y. Moriguchi, E. Rodenburg, D. Rogich, and H. Schütz. 1997. *Resource Flows: The Material Base of Industrial Economies*. Washington, DC: World Resources Institute.
Ahmed, A. and M. Ramzan. 2013. Effects of job stress on employee's job performance: A study on banking sector of Pakistan. *IOSR Journal of Business and Management, 11*(6): 61–68.
Akram, K., S. H. Siddiqui, M. A. Nawaz, T. A. Ghauri, and A. K. H. Cheema. 2011. Role of knowledge management to bring innovation: An integrated approach. *International Bulletin of Business Administration, 11*: 121–134.
Alberoni, F. 2006. *Menas vadovauti*. Vilnius: Dialogo kultūros centras.
Ali, I., K. U. Rehman, S. I. Ali, J. Yousaf, and M. Zia. 2010. Corporate social responsibility influences, employee commitment and organisational performance. *African Journal of Business Management, 4*(13): 2796–2801.
Alisauskas, A., S. Alisauskiene, D. Gerulaitis, R. Meliene, L. Milteniene, O. Sapelyte. 2007. Level and need of special pedagogical assistance in Lithuanian general education schools. Research Report. Siauliai University Printing House.
Alvarez, S. A., R. D. Ireland, and J. J. Reuer. 2006. Entrepreneurship and strategic alliances. *Journal of Business Venturing, 21*(4): 401–404.
Amar, A. D. and C. Hentrich. 2009. To be a better leader, give up authority. *Harvard Business Review, 87*(12): 22–24.
Amatori, F. and A. Colli. 2013. *Business History: Complexities and Comparisons*. 1st Edition. London: Routledge.
Anand, P. 2016. Happiness, well-being and human development: The case for subjective measures. *UNDP Human Development Report*, 36.
Anderberg, S., S. Prieler, K. Olendrzynski, and S. de Bruyn. 2000. *Old Sins: Industrial Metabolism, Heavy Metal Pollution and Environmental Transition in Central Europe*. Tokyo: UN University Press.
Andersson, M., B. Johansson, C. Karlsson, and H. Lööf (Eds.). 2012. *Innovation and Growth. From R&D Strategies of Innovating Firms to Economy-Wide Technological Change*. Oxford: Oxford University Press.
Ansoff, I. 1957. Strategies for diversification. *Harvard Business Review, 35*(5): 113–124.
Ardichvili, A. and S. Manderscheid. 2008. Emerging practices in leadership development: An introduction. *Advances in Developing Human Resources, 10*(5): 619–631.

Atkociuniene, Z. O. and R. Radiunaite. 2011. Žinių vadybos įtaka darnaus vystymosi reikšmėms įgyvendinti organisacijoje. *Informacijos mokslai*, Vilnius, *58*: 56–73.

Avolio, B. and W. Gardner. 2005. Authentic leadership development: Getting to the root of positive forms of leadership. *Leadership Quarterly*, *16*(3): 315–338.

Avolio, B. J., W. L. Gardner, F. Luthans, and F. O. Walumbwa. 2005. Authentic leadership development: Building integration and differentiation. *Leadership Quarterly*, *16*(3): 343–372.

Ayres, R. U., L. W. Ayres, and B. Warr. 2003. Exergy, power and work in the US economy, 1900–1998. *Energy*, *28*(3): 219–273.

Bagdoniene, D., A. Galbuogiene, and E. Paulaviciene. 2009. Formation of sustainable organization concept on the basis of global quality management. *Economics and Management*, *14*. Kaunas: KTU.

Bailur, S. 2006. Using stakeholder theory to analyze telecenter projects. *Information Technologies and International Development*, *3*(3): 61–80.

Baldassarri, S. and T. Saavala, T. 2006. Entrepreneurship—educating the next generation of entrepreneurs. *Enterprise Europe*, *22*: 16–20.

Bandzeviciene, R. 2011. Inovacijų vadybos psichologija. Mykolo Romerio universitetas, Vilnius.

Banerjee, K., C. F. Chabris, V. E. Johnson, J. J. Lee, F. Tsao, and M. D. Hauser. 2009. General intelligence in another primate: Individual differences across cognitive task performance in a new world monkey (Saguinus oedipus). *PloS One*, *4*(6): 58–83.

Baron, R. 2012. *Entrepreneurship: An Evidence-based Guide*. Cheltenham: Edward Elgar Publishing.

Barrett, R. 2003. Culture and consciousness: Measuring spirituality in the workplace by mapping values. In R. A. Giacalone and C. L. Jurkiewicz (Eds.), *Handbook of Workplace Spirituality and Organisational Performance* (pp. 345–366). New York: M. E. Sharp.

Bartelmus, P. 2013. The future we want: Green growth or sustainable development? *Environmental Development*, *7*: 165–170.

Barth, M. and A. Busch. 2006. Competencies and higher education for sustainable development. Working paper. Institute for Environmental and Sustainability Communication, University of Lüneburg, Lüneburg.

Barzdaite, A. and R. Navickiene. 2013. Expression of entrepreneurship in tourism business. *Science and Practice: Topicalities and Perspectives*. A collection of scientific articles. Kaunas, 26–36.

Bass, S. and B. Dalal-Clayton. 2002. *Sustainable Development Strategies: A Resource Book*. New York: Earthscan.

Basu, K. and G. Palazzo. 2008. Corporate social responsibility: A process model of sensemaking. *Academy of Management Review*, *33*(1): 122–136.

Batista, A. A. S. and A. C. Francisco. 2018. Organisational sustainability practices: A study of the firms listed by the Corporate Sustainability Index. *Sustainability*, *10*(1): 226.

Bereiter, C. and M. Scardamalia. 2003. Learning to work creatively with knowledge. In E. De Corte, L. Verschaffel, N. Entwistle, and J. van Merriënboer (Eds.), *Powerful Learning Environments. Unraveling Basic Components and Dimensions* (pp. 55–68; Advances in Learning and Instruction Series). Oxford: Elsevier Science.

Bernhardt, A. and P. Osterman. 2017. Organising for good jobs: Recent developments and new challenges. *Work and Occupations*, *44*(1): 89–112.

Bethel, S. 2004. *Value Based Leadership Essentials for the 21 Century*. www.bethelinstitute.com (accessed July 24, 2019).

Bhattacharya, C. B., S. Sen, and D. Korschun. 2008. Using corporate social responsibility to win the war for talent. *MIT Sloan Management Review*, *49*(2): 37–44.

Biddle, J. E. and D. S. Hamermesh. 1998. Beauty, productivity and discrimination: Lawyers' looks and lucre. *Journal of Labor Economics*, *16*: 172–201.

Bigliardi, B. and A. I. Dormio. 2009. An empirical investigation of innovation determinants in food machinery enterprises. *European Journal of Innovation Management, 12*(2): 223–242.

Bird, R., A. D. Hall, F. Momente, and F. Reggiani. 2007. What corporate social responsibility activities are valued by the market? *Journal of Business Ethics, 76*(2): 189–206.

Birth, G., L. Illia, F. Lurati, and A. Zamparini. 2008. Communicating CSR: practices among Switzerland's top 300 companies. *Corporate Communications: An International Journal, 13*(2): 182–196.

Biswas, S. 2009. Organisation culture and transformational leadership as predictors of employee performance. *The Indian Journal of industrial Relations, 44*(4): 611–627.

Blackwell, S. 2006. The influence of perceptions of organisational structure and culture on leadership role requirements: The moderating impact of locus of control and self-monitoring. *Journal of Leadership and Organisational Studies, 12*(4): 1–27.

Blake, R. R. and J. S. Mouton. 1966. *Managerial Grid: Leadership Styles for Achieving Production Through People*. 9th Edition. Houston, TX: Gulf Publishing.

Blomback, A. and C. Wigren. 2009. Challenging the importance of size as determinant for CSR activities. *Management of Environmental Quality: An International Journal, 30*(3): 256–270.

Boehe, D. M. and L. B. Cruz. 2010. Corporate social responsibility, product differentiation strategy and export performance. *Journal of Business Ethics, 91*: 325–346.

Boehe, D. M., L. B. Cruz, and M. H. Ogasavara. 2010. How can firms from emerging economies enhance their CSR-supported export strategies? Insper, Ibmec, São Paulo.

Bogdanov, E. N. and V. G. Zazykin. 2003. *Psihologicheskie osnovy `Pablik rileyshnz`*. St Petersburg: Piter.

Bolden, R. 2007. Trends and perspectives in management and leadership. *Business Leadership Review IV: II*, April 2007.

Bolton, B. and J. Thompson. 2004. *Entrepreneurs: Talent, Temperament, Technique*. London: Routledge.

Boons, F. and F. Ludeke-Freund. 2013. Business models for sustainable innovation: State-of-the-art and steps towards a research agenda. *Journal of Cleaner Production, 45*: 9–19.

Boulouta, I. and C. N. Pitelis. 2014. Who needs CSR? The impact of corporate social responsibility on national competitiveness, *Journal of Business Ethics, 119*(3): 349–364.

Boxall, P. F., J. Purcell, and P. Wright. 2007 Human resource management; scope, analysis and significance. In P. Boxall, J. Purcell, and P. Wright (Eds.), *The Oxford Handbook of Human Resource Management* (pp. 1–18). Oxford: Oxford University Press.

Boyatzis, R. and A. McKee. 2005. *Resonant Leadership: Renewing Yourself and Connecting with Others Through Mindfulness, Hope, and Compassion*. Boston, MA: Harvard Business School Press.

Boyatzis, R. and A. McKee. 2006. *Sustainable Leadership*. Vilnius: Verslo zinios.

Boyd, B. K., D. D. Bergh, and D. J. Ketchen Jr. 2010. Reconsidering the reputation—performance relationship: A resource-based view. *Journal of Management, 36*(3): 588–609.

Brammer, S., G. Jackson, and D. Matten. 2012. Corporate social responsibility and institutional theory: New perspectives on private governance. *Socio-Economic Review, 10*(1): 3–28.

Bruyn, S. M. 2000. *Environmental Growth and the Environment: An Empirical Analysis*. Dordrecht: Kluwer Academic Publishers.

Buchen, I. H. 1998. Servant leadership: A model for future faculty and future institutions. *Journal of Leadership Studies, 5*: 125.

Buford, B. A. 2001. Management effectiveness, personality, leadership, and emotional intelligence: A study of the validity evidence of the Emotional Quotient Inventory (EQ-i). PhD dissertation, University of Iowa, IA.

Bullock, S. 2001. Keeping score: In search of an accurate Environmental Sustainability Index. www.onlineopinion.com.au/view.asp?article=1213 (accessed July 24, 2019).

Burkart, J. M., M. N. Schubiger, and C. P. van Schaik. 2016. The evolution of general intelligence. *Behavioral and Brain Sciences, 1*: 65.

Burlea-Schiopoiu, A. 2013. Global environmental management initiative. In S. O. Idowu, N. Capaldi, L. Zu, and A. Das Gupta (Eds.), *Encyclopedia of Corporate Social Responsibility* (pp. 1241–1248). Berlin, Heidelberg: Springer-Verlag.

Busenitz, L. W. and C. M. Lau. 1997. A cross-cultural cognitive model of new venture creation. *Entrepreneurship Theory and Practice, 20*(4): 25–40.

Caldwell, S. D., D. M. Herold, and D. B. Fedor. 2004. Toward an understanding of the relationships among organisational change, individual differences, and changes in person-environment fit: A cross-level study. *Journal of Applied Psychology, 89*(5): 868–882.

Cameron, K. S., R. E. Quinn, J. De Graff, and A. V. Thakor. 2006. *Competing Values Leadership: Creating Value in Organisations*. Cheltenham: Edward Elgar.

Cardona, P. and C. Rey. 2008. *Management by Missions*. Basingstoke: Palgrave Macmillan.

Carnegie, D. 2011. *The Art of Leadership. How to Encourage Yourself and Others to Strive for Perfection*. Vilnius: Eugrimas.

Carré, F., P. Findlay, C. Tilly, and C. Warhurst. 2012. Job quality: Scenarios, analysis and interventions. In C. Warhurst, P. Findlay, C. Tilly, and F. Carré (Eds.), *Are Bad Jobs Inevitable? Trends, Determinants and Responses to Job Quality in the Twenty-First Century*. London: Palgrave Macmillan.

Carrillo-Hermosilla, J., P. Del Río, and T. Könnölä. 2010. Diversity of eco-innovations: Reflections from selected case studies. *Journal of Cleaner Production, 18*(10–11): 1073–1083.

Carsrud, A. L. and M. E. Brännback. 2007. *Entrepreneurship*. Westport, CT; London: Greenwood Publishing Group.

Casson, M. 2003. *The Entrepreneur: An Economic Theory. Modern Revivals in Economics*. Cheltenham: Edward Elgar Publishing.

Castro, R. G. 2013. The relationship of leadership styles and organisational productivity skills to teacher professionalism. *Journal of Education and Practice, 4*(7): 135–142.

Centre for Economics and Business Research. 2014. *The Future Economic and Environmental Costs of Gridlock in 2030*. London: CEBR.

Charter, M. and T. Clark. 2007. *Sustainable Innovation*. Farnham: The Centre for Sustainable Design.

Charter, M., C. Gray, T. Clark, and T. Woolman. 2008. Review: The role of business in realising sustainable consumption and production. In A. Tukker, M. Charter, C. Vezzoli, E. Stø, and M. M. Andersen (Eds.), *Perspectives on Radical Changes to Sustainable Consumption and Production. System Innovation for Sustainability* (pp. 46–69). Sheffield: Greenleaf.

Chlivickas, E., N. Petrauskaite, and N. Ambrusevic. 2009. Leading priorities for development of the high technologies market. *Journal of Business Economics and Management, 10*(4): 321–328.

Chopra, P. K. and G. K. Kanji. 2010. Emotional intelligence: A catalyst for inspirational leadership and management excellence. *Total Quality Management, 21*(10): 971–1004.

Christou, E., S. Avdimiotis, P. Kassianidis, and M. Sigala. 2004. Examining the factors influencing the adoption of web-based ticketing: Etix and its adopters. Information and Communication Technologies in Tourism 2004, the 11th ENTER International Conference in Cairo, Egypt, 2004: 129–138.

Ciegis, R. 2004. *Economy and Environment: Managing Sustainable Development. Monograph*. Kaunas: Vytautas Magnus University.

Ciegis, R. 2006. Ecological security: New challenges for the planet. *Strategic Self-Management, 1*(3): 22–33. Klaipėda: Vilnius University Press.

Ciegis, R. and R. Grunda. 2007. The process of transforming a company into a sustainable company. *Organizational Management: Systematic Research, 44*: 19–34.

Ciegis, R. and J. Ramanauskiene. 2011. Integruotas darnaus vystymosi vertinimas: Lietuvos atvejis. *Management Theory and Studies for Rural Business and Infrastructure Development*, *2*(26): 39–49.

Ciegis, R., J. Ramanauskiene, and B. Martinkus. 2009. The concept of sustainable development and its use for sustainability scenarios. *Engineering Economics*, *2*(62): 28–37.

Ciegis, R., J. Ramanauskiene, and G. Startiene. 2009. Theoretical reasoning of the use of indicators and indices for sustainable development assessment. *Engineering Economics*, *3*(63): 33–40.

Ciegis, R. and R. Zeleniutė. 2008. Economic development in terms of sustainable development. *Applied Economics: Systematic Research*, *2*(1): 39–53.

Cipresso, P., S. Serino, and G. Riva. 2014. The pursuit of happiness measurement: A psychometric model based on psychophysiological correlates. *Scientific World Journal*, *2014*, Article ID 139128. http://dx.doi.org/10.1155/2014/139128 (accessed July 24, 2019).

CISL. 2011. *A Journey of a Thousand Miles: The State of Sustainability Leadership.* Cambridge: Cambridge Institute for Sustainability Leadership.

Clarkson, M. B. E. 1995. A stakeholder framework for analyzing and evaluating corporate social performance. *Academy of Management Review*, *20*(1): 92–117.

Clifton, D. O. and J. K. Harter. 2003. Investing in strengths. In K. S. Cameron, J. E. Dutton, and R. E. Quinn (Eds.), *Positive Organisational Scholarship* (pp. 111–121). San Francisco, CA: Berret-Koehler.

Collier, J. and R. Esteban. 2007. Corporate social responsibility and employee commitment. *Business Ethics: A European Review*, *16*(1): 19–33.

Collins, V. L. 2001. Emotional intelligence and leader success. PhD dissertation, University of Nebraska, Lincoln, NE (Publication No. AAT 3034371).

Connelly, B. L., R. D. Ireland, C. R. Reutzel, and J. E. Coombs. 2010. The power and effects of entrepreneurship research. *Entrepreneurship Theory and Practice*, *34*(1): 131–149.

Cooke, P., M. Heidenreich, and H. J. Braczyk. 2004. *Regional Innovation Systems*. London, New York: Routledge.

Cooper, C. L. 2005. *Leadership and Management in the 21st Century: Business Challenges of the Future*. Oxford: Oxford University Press.

Cooper, C., T. Scandura, and C. Schriesheim. 2005. Looking forward but learning from the past: Potential challenges to developing authentic leadership theory and authentic leaders. *Leadership Quarterly*, *16*(3): 475–493.

Cordona, P. and C. Rey. 2014. *Mission-Based Management*. Vilnius: Vaga.

Crane, A. and D. Matten. 2007. *Business ethics: Managing corporate citizenship and sustainability in the age of globalization*. 2nd Edition. Oxford: Oxford University Press.

Crofton, F. S. 2000. Educating for sustainability: Opportunities in undergraduate engineering. *Journal of Cleaner Production*, *8*: 397–405.

Cummins, R. A. 2005. Moving from the quality of life concept to a theory. *Journal of Intellectual Disability Research*, *49*: 699–706.

Dahlsrud, A. 2008. How corporate social responsibility is defined: An analysis of 37 definitions. *Corporate Social Responsibility and Environmental Management*, *15*(1): 1–13.

De Bussy, N. M. and L. Suprawan. 2012. Most valuable stakeholders: The impact of employee orientation on corporate financial performance. *Public Relations Review*, *38*: 280–287.

De Medeiros, J. F., J. Ribeiro, and M. N. Cortimiglia. 2014. Success factors for environmentally sustainable product innovation: A systematic literature review. *Journal of Cleaner Production*, *65*: 76–86.

Dearing, J. A., R. Wang, K. Zhang, J. G. Dyke, H. Haberl, M. S. Hossain, P. G. Langdon, T. M. Lenton, K. Raworth, and S. Brown. 2014. Safe and just operating spaces for regional social-ecological systems. *Global Environmental Change*, *28*: 227–238.

Demirbas, D., J. G. Hussain, and H. Matlay. 2011. Owner–managers' perceptions of barriers to innovation: Empirical evidence from Turkish SMEs. *Journal of Small Business and Enterprise Development*, *18*(4): 764–780.

Derry, R., and R. M. Green. 1989. Ethical theory in business ethics: A critical assessment. *Journal of Business Ethics*, *8*(7): 521–533.

Dewe, P. 2002. *Stress. A brief history*. Oxford: Blackwell Publishing.

Dhammika, K. A. S. 2016. Visionary leadership and organisational commitment: The mediating effect of leader-member exchange (LMX). *Wayamba Journal of Management*, *4*(1): 1–10.

Dhang, S. 2012. The role of teachers in nation-building. *DYPIMS's International Journal of Management and Research*, *1*(1): 53–59.

Diamond, M. A. and S. Allcorn. 2009. *Private Selves in Public Organisations: The Psychodynamics of Organisational Diagnosis and Change*. New York: Palgrave Macmillan.

Dickinson-Delaporte, S., M. Beverland, and A. Lindgreen. 2010. Building corporate reputation with stakeholders. *European Journal of Marketing*, *44*(11/12): 1856–1874.

Diener, E. and S. Oishi. 2000. Money and happiness: Income and subjective well-being across nations. www.researchgate.net/publication/246865747 (accessed July 22, 2019).

Dixon, P. M., J. Weiner, T. Mitchell-Olds, and R. Woodley. 1988. Erratum to "Bootstrapping the Gini Coefficient of Inequality." *Ecology*, *69*: 1307.

Doane, D. 2005. Beyond corporate social responsibility: Minnows, mammoths and markets. *Futures*, *37*: 215–229.

Dodgson, M., D. Gann, and A. Salter. 2008. *The Management of Technological Innovation: Strategy and Practice*. Oxford: Oxford University Press.

Donaldson, T. and L. E. Preston. 1995. The stakeholder theory of the corporation: Concepts, evidence, and implications. *Academy of Management Review*, *20*(1): 65–91.

Donnison, P. 2008. Executive coaching across cultural boundaries: An interesting challenge facing coaches today. *Development and Learning in Organisations*, *22*(4): 17–19.

Doppelt, B. 2010. *Leading Change Toward Sustainability: A Change-Management Guide for Business, Government and Civil Society*. 2nd Edition. Sheffield: Greenleaf.

Dressler, S. 2004. *Strategy, Organisational Effectiveness and Performance Management: From Basics to Best Practices*. Boca Raton, FL: Universal Publishers.

Drucker, P. F. 2007. *The Practice of Management*. Oxford: Butterworth Heinemann.

Du, S., C. B. Bhattacharya, and S. Sen. 2011. Corporate social responsibility and competitive advantage: Overcoming the trust barrier. *Management Science*, *57*(9): 1528–1545.

Duknytė, R. 2015. *Leadership. Concepts. Classifications. Symptoms*. Vilnius: Lithuanian University of Educational Sciences.

Dulewicz, V. and M. Higgs. 2000. Emotional intelligence: A review and evaluation study. *Journal of Managerial Psychology*, *15*(4): 341–372.

Dutot, V., E. Lacalle Galvez, and D. W. Versailles. 2016). CSR communications strategies through social media and influence on e-reputation, *Management Decision*, *54*(2): 363–389.

Eagly, A. H. and J. L. Chin. 2010. Diversity and leadership in a changing world. *American Psychologist*, *65*(3): 216–224.

Easterlin, R. A., L. A. McVey, M. Switek, O. Sawangfa, and J. S. Zweig. 2011. The happiness-income paradox revisited. IZA Discussion Paper No. 5799.

Edmonds, S. C. 2014. *The Culture Engine: A Framework for Driving Results, Inspiring Your Employees, and Transforming Your Workplace*. Hoboken, NJ: Wiley.

Edwards, G., S. Turnbull, D. Stephens, and A. Johnston. 2008. *Developing Leadership for Sustainable Development*. Leadership: Impact, Culture and Sustainability Building Bridges Series. Silver Spring, MD: The International Leadership Association.

Ekelund, H. and Hébert, R. F. 2013. *A History of Economic Theory and Method*. 6th Edition. Long Grove, IL: Waveland Press.

Elkington, J. 2010. Corporate sustainability. In W. Visser, D. Matten, M. Pohl, and N. Tolhurst (Eds.), *The A to Z of Corporate Social Responsibility* (1st Edition; pp. 114–119). Chichester: Wiley Publishing.

Elliott, L. 2005. The United Nations' record on environmental governance: An assessment. In F. Biermann and S. Bauer (Eds.), *A World Environment Organisation: Solution or Threat for Effective International Environmental Governance?* (pp. 27–56). Aldershot: Ashgate.

Elloy, D. F. 2005. The influence of super leader behaviours on organisation commitment, job satisfaction and organisational self-esteem in a self-managed work team. *Leadership and Organisational Development Journal*, *26*(2): 120–127.

Elzen, B., F. W. Geels, and K. Green. 2005. *System Innovation and the Transition to Sustainability: Theory, Evidence and Policy*. Cheltenham: Edward Elgar Publishing.

Ensley, M. D., K. M. Hmieleski, and C. L. Pearce. 2006. The importance of vertical and shared leadership within new venture top management teams: Implications for the performance of startups. *The Leadership Quarterly*, *17*: 217–231.

Erhel, C. and M. Guergoatlarivicre. 2010. Job quality and labour market performance. CEPS Working Document No 330, June. Centre for European Policy Studies.

Esigbone, E. M. 2000. Influence of perceived leadership style on employee's job satisfaction. BSc research project, University of Lagos, Nigeria.

Essien, E. A., O. A. Adekunle, and A. M. Oke Bello. 2013. Management style and staff turnover in Nigerian banks: A comparative analysis: *American International Journal of Social Science*, *2*(6): 79–93.

European Commission. 2004. *Innovation Management and the Knowledge-Driven Economy*. Brussels, Luxembourg: ECSC-EC-EAEC.

European Commission and Eurostat. 1999. *Towards Environmental Pressure Indicators for the EU*. 1st Edition. Brussels: European Commission and Eurostat.

European Commission and Eurostat. 2001. *Towards Environmental Pressure Indicators for the EU*. 2nd Edition. Brussels: European Commission and Eurostat.

Fagerberg, J., M. Srholec, and B. Verspagen. 2009. Innovation and economic development. In B. Hall and N. Rosenberg (Eds.), *Handbook of the Economics of Innovation* (Vol. II, p33–872). North Holland: Elsevier.

Ferdig, M. 2007. Sustainability leadership: Co-creating a sustainable future. *Journal of Change Management*, *7*(1): 25–35.

Fifield, L. K. 2008. Accelerator mass spectrometry of the actinides. *Quaternary Geochronology*, *3*(3): 276–290.

Flowers, B. S. 2008. *CISL Sustainability Leadership Research Interview*. Cambridge Institute for Sustainability Leadership and W. Visser.

Forsythe, S., M. Drake, and C. Cox. 1985. Influence of applicant's dress on interviewers' selection decisions. *Journal of Applied Psychology*, *70*: 374–378.

Freeman, R. E., J. S. Harrison, and A. Wicks. 2008. *Managing for Stakeholders: Survival, Reputation, and Success*. New Haven, CT: Yale University Press.

Frostenson, M., S. Helin, and J. Sandström. 2011. Organising corporate responsibility communication through filtration: A study of web communication patterns in Swedish retail. *Journal of Business Ethics*, *100*(1): 31–43.

Fry, W. L. 2003. Toward a theory of spiritual leadership. *The Leadership Quarterly*, *14*: 693–727.

Fry, W. L., L. L. Matherly, J. L. Whittington, and B. E. Winston. 2007. Spiritual leadership as an integrating paradigm for servant leadership. In S. Singh-Sengupta and D. Fields (Eds.), *Integrating Spirituality and Organisational Leadership* (pp. 70–82). New Delhi, India: Macmillan India, Ltd.

Gallaher, M., A. Link, and J. Petrusa. 2007. *Marshall and Schumpeter on Evolution: Economic Sociology of Capitalist Development*. Cheltenham: Edward Elgar Publishing.

Gammoh, B. and K. E. Voss. 2011. Brand alliance research: In search of a new perspective and directions for future research. *Journal of Marketing Development and Competitiveness*, *5*(3): 81–93.

Garriga, E. and D. Melé. 2004. Corporate social responsibility theories: Mapping the territory. *Journal of Business Ethics*, *53*: 51–71.

Gartner, W. B. 1990. What about when we talk about entrepreneurship? *Journal of Business Venturing*, *1*: 15–28.

Gehin, A., P. Zwolinski, and D. Brissaud. 2008. A tool to implement sustainable end-of-life strategies in the product development phase. *Journal of Cleaner Production*, *16*: 566–576.

Godfrey, P. C., C. B. Merrill, and J. M. Hansen. 2009. The relationship between corporate social responsibility and shareholder value: An empirical test of the risk management hypothesis. *Strategic Management Journal*, *30*(4): 425–445.

Goffee, R. and G. Jones. 2009. Authentic leadership. *Leadership Excellence*, May.

Gold, J., R. Thorpe, and A. Mumford. 2010. *Handbook of Leadership and Management Development*. 5th Edition. Farnham: Gower Publishing Ltd.

Goldsmith, M. and L. S. Lyons. 2006. *Coaching for Leadership: The Practice of Leadership Coaching from the World's Greatest Coaches*. San Francisco, CA: Pfeiffer.

Goleman, D. 1995. *Emotional Intelligence*. New York: Bantam Books.

Goleman, D. 1997. Developing the leadership competencies of emotional intelligence. Presented at the 2nd International Competency Conference, London, October 1997.

Goleman, D. 1998. *Working with Emotional Intelligence*. New York: Bantam Books.

Goleman, D. 2000. Emotional intelligence: Issues in paradigm building. In D. Goleman, and C. Cherniss (Eds.), *The Emotionally Intelligent Workplace: How to Select for, Measure, and Improve Emotional Intelligence in Individuals, Groups, and Organisations*. San Francisco, CA: Jossey-Bass.

Goleman, D., R. Boyatzis, and A. McKee. 2002. *The New Leaders: Transforming the Art of Leadership into the Science of Results*. London: Time-Warner.

Goleman, D., R. Boyatzis, and A. McKee. 2007. *Leadership: How to Lead Through Emotional Intelligence*. Kaunas: Smaltija.

Gottfredson, L. S. 1998. The general intelligence factor. *Scientific American Presents*, *9*: 24–30.

Goudarzvand-Chegini, M. and S. G. Mirdoozandeh. 2012. Relationship between quality of work-life and job satisfaction of the employees in public hospitals in Rasht. *Zahedan Journal of Research in Medical Sciences*, *14*(2): 108–111.

Graham, A. 2005. *Supercoaching: The Missing Ingredient for High Performance*. London: Random House Business.

Gräuler, M. and F. Teuteberg. 2014. *Greenwashing in Sustainability Communication: A Quantitative Investigation of Trust-Building Factors*. Conference: Environmental and Sustainability Management Accounting Network (EMAN), Rotterdam.

Griggs, D., M. Stafford-Smith, O. Gaffney, J. Rockström, M. C. Ohman, P. Shyamsundar, W. Steffen, G. Glaser, N. Kanie, and I. Noble. 2013. Policy: Sustainable Development Goals for people and planet. *Nature*, *495*: 305–307.

Grossman, G. M. and A. B. Krueger. 1991. Environmental impact of a North American Free Trade Agreement. Working paper 3914. Cambridge, MA: National Bureau of Economic Research.

Hagedoorn, J. 1996. Innovation and entrepreneurship: Schumpeter revisited. *Industrial and Corporate Change*, *5*(3): 883–896.

Hajer, M. A. 2009. *Authoritative Governance: Policy Making in the Age of Mediatization*. Oxford: Oxford University Press.

Hajer, M. A. 2011. *The Energetic Society. In Search of a Governance Philosophy for a Clean Economy.* The Hague, Netherlands: PBL Netherlands Environmental Assessment Agency.

Halford, J. T. and S. Hsu. 2014. Beauty is wealth: CEO appearance and shareholder value (December 19). https://ssrn.com/abstract=2357756; http://dx.doi.org/10.2139/ssrn.2357756.

Hall, P. A. and D. Soskice. 2001. *Varieties of Capitalism: The Institutional Foundations of Comparative Advantage*. Oxford: Oxford University Press.

Hamermesh, D. S. 2011. *Beauty Pays: Why Attractive People are More Successful*. Princeton, NJ: Princeton University Press.

Hamermesh, D. S. and J. E. Biddle. 1994. Beauty and the labor market. *The American Economic Review, 84*(5): 1174–1194.

Hamermesh, D. S., X. Meng, and J. Zhuang. 2001. Dress for success – does primping pay? *NBER Working Paper*, No. w7167.

Hamermesh, D. S. and A. Parker. 2005. Beauty in the classroom: Instructors' pulchritude and putative pedagogical productivity. *Economics of Education Review, 24*(4): 369–376.

Haque, M. M. and M. A. Taher. 2008. Job characteristics model and job satisfaction: Age, gender and marital status effect. Paper presented at the 7th International Conference on Ethics and Quality of Work-life for Sustainable Development, Bangkok, Thailand.

Harger, J. A. E. and F. M. Meyer. 1996. Definition of indicators for environmentally sustainable development. *Chemosphere, 33*: 1749–1775.

Hargreaves, A. and Fink, D. 2004. The seven principles of sustainable leadership. *Educational Leadership: Journal of the Department of Supervision and Curriculum Development*, NEA, *61*(7): 8–13.

Hargreaves, A. and Fink, D. 2006. *Sustainable Leadership*. San Francisco, CA: Jossey-Bass.

Haris, A. 2010. *Shared Leadership. Developing Future Leaders*. Vilnius: Education Supply Center.

Harper, B. 2000. Beauty, stature and the labour market: A British cohort study. *Oxford Bulletin of Economics and Statistics, 62*: 771–800.

Hart, V., Blattner, J., and S. Leipsic. 2001. Coaching versus therapy: A perspective. *Consulting Psychology Journal: Practice & Research, 53*(4): 229–237.

Helliwell, J. F., R. Layard, and J. D. Sachs. 2018. *The World Happiness Report*. New York: Sustainable Development Solutions Network. http://worldhappiness.report/ (accessed July 23, 2019).

Herendeen, R. 2004. Energy analysis and emergy analysis—a comparison. *Ecological Modelling, 178*: 227–237.

Heskett, J. 2007. How much of leadership is about control, delegation, or theater. *Harvard Business School*. http://hbswk.hbs.edu/item/5718.html (accessed July 23, 2019).

Hisrich, R. D. 1986. *Entrepreneurship, Intrapreneurship, and Venture Capital: The Foundation of Economic Renaissance*. Lexington, MA: Lexington Books.

Hisrich, R. D., M. P. Peters, and D. A. Shepherd. 2006. *Entrepreneurship*. Burr Ridge, IL: Irwin Professional Pub.

Hitt, M. A., R. D. Ireland, S. M. Camp, and D. L. Sexton. 2001. Strategic entrepreneurship: Entrepreneurial strategies for wealth creation. *Strategic Management Journal, 22*(6): 479–492.

Hofstede, G. 1980. *Culture's Consequences: International Differences in Work-Related Values (Vol. 5)*. Thousand Oaks, CA: Sage Publications, Incorporated.

Holmberg, J. and K-H. Robèrt. 2000. Backcasting from non-overlapping sustainability principles—a framework for strategic planning. *International Journal of Sustainable Development and World Ecology, 7*: 1–18.

Hopwood, B., M. Mellor, and G. O'Brien. 2005. Sustainable development: Mapping different approaches. *Sustainable Development, 13*: 38–52.

Horizon 2020. *Societal Challenge 5: Climate, Environment, Resource Efficiency and Raw Materials. 2014–2015 Work Program and Calls for Proposals*. Brussels: European Commission.

House, R. J., P. W. Dorfman, M. Javidan, P. J. Hanges, and M. F. Sully de Luque. 2014. *Strategic Leadership Across Cultures. GLOBE Study of CEO Leadership Behavior and Effectiveness in 24 Countries*. Thousand Oaks, CA: Sage Publications, Inc.

House, R. J. and J. M. Howell. 1992. Personality and charismatic leadership. *Leadership Quarterly, 3*: 81–108.

Hovelius, K. 1997. *Energy-, Exergy- and Emergy Analysis of Biomass Production*, Report, vol. 222, Swedish University of Agricultural Sciences, Uppsala.

HRM Guide. 2005. Appearance-based discrimination. www.hrmguide.com/diversity/appearance-at-work.htm

Hudson, F. M. 1999. *The Handbook of Coaching: A Comprehensive Resource Guide for Managers, Executives, Consultants, and Human Resource Professionals*. San Francisco, CA: Jossey-Bass.

Husted, B. and D. Allen. 2006. Corporate social responsibility in the multinational enterprise: Strategic and institutional approaches. *Journal of International Business Studies, 37*(6): 838–849.

Ibarra, H. 2012. Is "command and collaborate" the new leadership model? http://blogs.hbr.org/cs/2012/02/is_command_and_collaborate_the.html (accessed July 23, 2019).

Inglehart, R. and H. D. Klingemann. 2000. Genes, culture, democracy, and happiness. In E. Diener and E. M. Suh (Eds.), *Culture and Subjective Well-Being* (pp. 165–183). Cambridge, MA: MIT Press.

Ireland, R. D. and M. A. Hitt. 1999. Achieving and maintaining strategic competitiveness in the 21st century: The role of strategic leadership. *Academy of Management Executive, 13*(1): 43–57.

Islam, Z. M. and S. Siengthai. 2009. Quality of work life and organisational performance: Empirical evidence from Dhaka export processing zone. Paper presented at the ILO Conference on Regulating for Decent Work, International Labour Office, Geneva.

Jaeggi, S. M., M. Buschkuehl, J. Jonides, and W. J. Perrig. 2008. Improving fluid intelligence with training on working memory. *Proceedings of the National Academy of Sciences USA, 105*(19): 6829–6833.

Jakubavicius, A., R. Strazdas, and K. Gecas. 2003. *Business Innovation: Processes, Support, Networking*. Vilnius: Lithuanian Innovation Center.

Jamali, D., P. Lund-Thomsen, and S. Jeppesen. 2015. SMEs and CSR in developing countries. *Business & Society, 56*(1): 1–12.

Jamali, D. and R. Mirshak. 2007. Corporate social responsibility (CSR): Theory and practice in a developing country context. *Journal of Business Ethics, 72*(3): 243–262.

Jamali, D. and B. Neville. 2011. Convergence versus divergence in CSR in developing countries: An embedded multi-layered institutional lens. *Journal of Business Ethics, 102*: 599–621.

Jaruzelski, B. and K. Dehoff. 2010. *The Global Innovation 1000: How the Top Innovators Keep Winning*, Winter 2010/Issue 61 (originally published by Booz & Company).

Jenkins, R. 2005. Globalization, corporate social responsibility and poverty. *International Affairs, 81*(3): 525–540.

Jones, J. W., B. N. Barge, B. D. Steffy, L. M. Fay, L. K. Kuntz, and L. J. Wuebker. 1988. Stress and medical malpractice: Organisational risk assessment and intervention. *Journal of Applied Psychology, 73*(4): 727–735.

Jucevicius, R. 2007. *Map of Lithuanian Industrial and Business Clusters*. Kaunas: KTU Institute of Business Strategy.

Juknys, R. 2008. *Sustainable Development in Lithuania—present and Prospects. Industrial Dialogue with the Environment*. Kaunas: Vytautas Magnus University Press.

Jung, D. I. 2001. Transformational and transactional leadership and their effects on creativity in groups. *Creativity Research Journal, 13*: 185–195.

Juniper, E. F. 2002. Can quality of life be quantified? *Clinical and Experimental Allergy Reviews, 2*: 57–60.

Juodaityte, A. 2003. *Socialization and Education in Childhood*. Vilnius: UAB Petro Offset.

Kajzar, P. and M. Kozubkova. 2007. Quality of work life and job satisfaction. *Life Quality Conditions in Societies Basing on Information: Proceedings, 2*: 289–295.

Kalaiarasi, V. and S. Sethuram. 2017. Literature review on organisation culture and its influence. *International Journal of Advanced Research in Engineering & Management, 3*(8): 9–14.

Kale, S. and D. Arditi. 2010. Innovation diffusion modeling in the construction industry. *Journal of Construction Engineering and Management*, *136*(3): 329–340.

Kampa-Kokesch, S. 2003. *Executive Coaching as an Individually Tailored Consultation Intervention: Does It Increase Leadership?* Ann Arbor, MI: UMI.

Kampa-Kokesch, S. and M. Z. Anderson. 2001. Executive coaching: A comprehensive review of the literature. *Consulting Psychology Journal: Practice and Research*, *53*(4): 205–228.

Kampa-Kokesch, S. and R. P. White. 2002. The effectiveness of executive coaching: What we know and what we still need to know. In R. L. Lowman (Ed.), *The Handbook of Organisational Consulting Psychology* (pp. 139–158). San Francisco, CA: Jossey–Bass.

Kanter, R. M. 1983. *The Change Masters: Innovation for Productivity in the American Corporation.* New York: Simon & Schuster.

Keating, M. 2004. *The Earth Summit's Agenda for Change: A Plain Language Version of Agenda 21 and the Other Rio Agreements*. Geneva: Centre for Our Common Future.

Kennedy, D. N., N. Makris, M. R. Herbert, T. Takahasni, and S. Caviness. 2002. Basic principles of MRI and morphometry studies of human brain development. *Development Science*, *5*(3): 268–278.

Kenny, B. and E. Reedy. 2007. The impact of organisational culture factors on innovation levels in SMEs: An empirical investigation. *The Irish Journal of Management*, *1*(1): 119–142.

Keulartz, J., M. Korthals, M. Schermer, and T. Swierstra. 2004. Pragmatism in progress: A reply to Radder, Colapietro and Pitt. *Techné*, *7*(3): 38–48.

Kim, H. R., M. Lee, H. T. Lee, and N. M. Kim. 2010. Corporate social responsibility and employee–company identification. *Journal of Business Ethics*, *95*(4): 557–569.

Kirby, A. D. 2002. Entrepreneurship education: Can business schools meet the challenge? International Council for Small Business 47th world conference, Puerto Rico.

Kleijn, R. 2001. Adding it all up: The sense and non-sense of bulk-MFA. *Journal of Industrial Ecology*, *4*(2): 7–8.

Korhonen, J. 2003. Should we measure corporate social responsibility? *Corporate Social Responsibility and Environmental Management*, *10*: 25–39.

Kotter, J. P. A. 1990. *Force for Change: How Leadership Differs from Management*. New York: Free Press.

Kotterman, J. 2006. Leadership vs management: What's the difference? *Journal for Quality & Participation*, *29*(2): 13–17.

Krajnc, D. and P. Glavic. 2005. A model for integrated assessment of sustainable development. *Resources, Conservation and Recycling*, *43*: 189–208.

Kühn, A.-L., M. Stiglbauer, and M. Fifka. 2015. Contents and determinants of corporate social responsibility website reporting in Sub-Saharan Africa—a seven-country study. *Business and Society*, online first.

Lacy, P., J. Keeble, R. McNamara, J. Rutqvist, K. Eckelle, T. Haglund, P. Buddemeier, M. Cui, A. Sharma, A. Copper, T. Senior, and C. Petterson. 2014. *Circular Advantage: Innovative Business Models and Technologies to Create Value in a World without Limits to Growth*. Dublin: Accenture.

Lammers, P. E. M. and A. J. Gilbert. 1999. *Towards Environmental Pressure Indicators for the EU: Indicator Definition*. Brussels: Eurostat.

Landström, L. and Lohrke, F. 2010. *Historical Foundations of Entrepreneurial Research*. Cheltenham: Edward Elgar Publishing.

Lawrence, P. R. 2010. *Driven to Lead: Good, Bad and Misguided Leadership*. San Francisco, CA: John Wiley and Sons.

Layard, R. 2003. Has social science a clue? Income and happiness: Rethinking economic policy. Lionel Robbins memorial lecture series, March 3–5, 2003, London. http://eprints.lse.ac.uk/47427/ (accessed July 23, 2019).

Layard, R. 2005. Happiness: Lessons from a new science. *Foreign Affairs (Council on Foreign Relations)*, *84*(6), doi: 10.2307/20031793.
Lecerf, M. A. 2012. Internationalization and innovation: The effects of a strategy mix on the economic performance of French SMEs. *International Business Research*, *5*(6): 1–13.
Lee, M. P. 2008. A review of the theories of corporate social responsibility: Its evolutionary path and the road ahead. *International Journal of Management Reviews*, *10(*1): 53–73.
Leithy W. E. 2017. Organisational culture and organisational performance. *International Journal of Economics & Management Sciences*, *6*(4): 442.
Lekecinskaite, L. and S. Lesinskiene. 2017. Children and youth under 20 suicides in Lithuania 2010–2015. *Public Health*, *2*(77): 74–80.
Lessem, R. 1986. *Enterprise Development*. Aldershot: Gower.
Lin, C. H., H. L. Yang, and D. Y. Liou. 2009. The impact of corporate social responsibility on financial performance: Evidence from business in Taiwan. *Technology in Society*, *31*(1): 56–63.
Linnenluecke, M. K. and A. Griffiths. 2010. Corporate sustainability and organisational culture. *Journal of World Business*, *45*: 357–366.
Lithuanian Department of Statistics. 2017. *Lithuania in Numbers*. Vilnius: Lithuanian Department of Statistics, 53.
Lithuanian Progress Report. 2015. *A Smart Society*. Vilnius: Government of the Republic of Lithuania, 50.
Lucas, A. W., Y. S. Cooper, and S. MacFarlane. 2008. *Exploring Necessity-Driven Entrepreneurship in a Peripheral Economy*. Dublin: Institute for Small Business and Entrepreneurship.
Lundvall, B. (Ed.). 1992. *National Systems of Innovation: Towards a Theory of Innovation and Interactive Learning*. London: Pinter Publishers.
Lundvall, B. and B. Johnson. 1994. The learning economy. *Journal of Industry Studies*, *1*(2): 23–42.
Madrid-Guijarro, A., D. Garcia-Perez-de-Lema, and H. Van Auken. 2009. Barriers to innovation among Spanish manufacturing SMEs. *Journal of Small Business Management*, *47*(4): 465–488.
Mamedaitytė, S. 2003. *Public Relations*. Kaunas: Authorized Methodological Material.
Marimon, F., I. Heras, and M. Casadesu's. 2009. ISO 9000 and ISO 14000 standards: A projection model for the decline phase. *Total Quality Management*, *20*(1): 1–21.
Martin, J. 2002. *Organisational Culture: Mapping the Terrain*. Thousand Oaks, CA: Sage.
Martínez, P. and I. R. del Bosque. 2013. CSR and customer loyalty: The roles of trust, customer identification with the company and satisfaction, *International Journal of Hospitality Management*, *35*: 89–99.
Matuleviciene, M. and J. Stravinskiene. 2015. The importance of stakeholders for corporate reputation. *Inzinerine Ekonomika [Engineering Economics]*, *26*(1): 75–83.
Mauerhofer, V. 2008. 3-D Sustainability: An approach for priority setting in situation of conflicting interests towards a sustainable development. *Ecological Economics*, *64*: 496–506.
McKinnon, A., M. Browne, A. Whiteing, and M. Piecyk. 2015. *Green Logistics: Improving the Environmental Sustainability of Logistics*. 3rd Edition. London: Kogan Page Limited.
McWilliams, A. and S. Donald. 2001. Corporate social responsibility: A theory of the firm perspective. *The Academy of Management Review*, *26*(1): 117–127.
McWilliams, A., D. S. Siegel, and P. M. Wright. 2006. Corporate social responsibility: Strategic implications. *Journal of Management Studies*, *43*(1): 2–17.
Melnikas, B., A. Jakubavičius, and R. Strazdas. 2000. *Innovation: Business, management, consulting*. Vilnius: Lithuanian Innovation Center.
Mester, C., D. Visser, and G. Roodt. 2003. Leadership style and its relation to employee attitudes and behaviour. *South African Journal of Industrial Psychology*, *29*(2): 72–82.

Mikalauskiene, A. and D. Streimikienė. 2014. *Theoretical Framework for Implementing Sustainable Development Policy*. Learning tool. Vilnius University.

Mori, K. and A. Christodoulou. 2012. Review of sustainability indices and indicators: Towards a new City Sustainability Index (CSI). *Environmental Impact Assessment Review, 32*: 94–106.

Morsing, M. and D. Oswald. 2009. Sustainable leadership: Management control systems and organisational culture in Novo Nordisk A/S. *Corporate Governance, 9*: 83–99.

Morsing, M. and M. Schultz. 2006. Corporate social responsibility communication: Stakeholder information, response and involvement strategies. *Business Ethics: A European Review, 15*(4): 323–338.

Moura-Leite, R. C. and R. C. Padgett. 2011. Historical background of corporate social responsibility. *Social Responsibility Journal, 7*(4): 528–539.

Muller-Camen, M., R. Croucher, and S. Leigh. 2008. Human *Resource Management: A Case Study Approach*. London: Chartered Institute of Personnel and Development.

Murphy, C. and J. Rosenfield. 2016. *The Circular Economy: Moving from Theory to Practice*. New York: McKinsey & Company Practice Publications.

Nathan, Joshua D. 2018. Is The Natural Step's theory about sustainability still sustainable? A theoretical review and critique. *Journal of Sustainable Development, 11*(1): 125–139.

National Environmental Strategy. 2015. UAB "ARX Reklama," Kaunas, p. 104.

National Strategy for Sustainable Development. 2009. Resolution no. 1247 Revision. Government of the Republic of Lithuania, September 16, p. 83.

Navajas, A., L. Uriarte, and Luis M. Gandía. 2017. Application of eco-design and life cycle assessment standards for environmental impact reduction of an industrial product. *Sustainability, 9*: 1724.

Nazali, M., M. Noor, and M. Pitt. 2009. A critical review on innovation in facilities management service delivery. *Facilities, 27*(5/6): 211–228.

Neely, A. and J. Hii. 1998. *Innovation and Business Performance: A Literature Review*. The Judge Institute of Management Studies, University of Cambridge, commissioned by GO-ER, January 15.

Neville, B. A., S. J. Bell, and B. Menguc. 2005. Corporate reputation, stakeholders and the social performance—financial performance relationship. *European Journal of Marketing, 39*(9/10): 1184–1198.

Newell, P. A. 2002. World Environmental Organisation: The wrong solution to the wrong problem. *World Economy, 25*: 659–671.

Nisbet. 1980. *History of the Idea of Progress*. London, New York: Transaction Publishers.

Noel, M. and M. Luckett. 2014. The benefits, satisfaction, and perceived value of small business membership in a chamber of commerce. *International Journal of Nonprofit and Voluntary Sector Marketing, 19*: 27–39.

Norman, W. 2013. Business ethics. In H. LaFollette (Ed.), *The International Encyclopedia of Ethics* (pp. 652–668). Chichester: Wiley-Blackwell.

Northouse, P. G. 2019. *Leadership: Theory and Practice*. 7th Edition. Thousand Oaks, CA: Sage Publications.

Nwadukwe, U. and O. T. Court. 2012. Management styles and organisational effectiveness: An appraisal of private enterprises in Eastern Nigeria. *American International Journal of Contemporary Research, 2*(9): 198–204.

Nykvist, B., Å. Persson, F. Moberg, L. Persson, S. Cornell, and J. Rockström. 2013. National environmental performance on planetary boundaries. In *Swedish Environmental Protection Agency Report 6576*, Swedish Environmental Protection Agency, Stockholm, Sweden.

OECD/Eurostat. 1997. *Oslo Manual: Guidelines for Collecting and Interpreting Innovation Data*. 3rd edition. Paris; OECD Publishing.

Ohme, E. 2002. *Guide for Managing Innovation. Part I: Diagnosis*. Barcelona: Centre for Innovation and Business Development.

Onetti, A., A. Zucchella, M. Jones, and P. McDougall-Covin. 2012. Internationalization, innovation and entrepreneurship: Business models for new technology-based firms. *Journal of Management & Governance*, *16*(3): 337–368.

O'Sullivan, D. and L. Dooley. 2009. *Applying Innovation*. Thousand Oaks, CA: Sage.

Pattberg, P. 2012. How climate change became a business risk: Analyzing nonstate agency in global climate politics. *Environment and Planning C: Government and Policy*, *30*: 613–626.

Pedersen, E. R. 2006. Making corporate social responsibility (CSR) operable: How companies translate stakeholder dialogue into practice. *Business and Society Review*, *111*(2): 137–163.

Peloza, J., M. Loock, J. Cerruti, and M. Muyot. 2012. Sustainability: How stakeholder perceptions differ from corporate reality. *California Management Review*, *55*(1): 74–97.

Pfahl, S. 2005. Institutional sustainability. *Sustainable Development*, *8*(1/2): 80–96.

Pfeifer, R. and J. C. Bongard. 2006. *How the Body Shapes the Way We Think: A New View of Intelligence*. Cambridge, MA: MIT Press.

Pikturniene, I. and J. Kurtinaitiene. 2010. *Consumer Behavior: Theory and Practice*. Vilnius: Vilnius University Press.

Pociute, B. 2005. Quality culture is a core value of academia. *Acta Paedagogica Vilnensia*, *15*: 188–196.

Pongrácz, E. 2006. Industrial ecology and waste management: From theories to applications. *Progress in Industrial Ecology: An International Journal*, *3*(1/2): 59–74.

Prendergast, R. 2011. And Shionoya. In J. E. Biddle and R. B. Emmett (Eds.), *Research in the History of Economic Thought and Methodology* (Volume 29, Part 1; pp. 175–184). Bingley: Emerald Group Publishing Limited.

Pusinaite, R. 2015. Sustainable innovation: Concept, ways of development, and development factors. In R. Pucetaite, A. Novelskaite, and R. Pušinaite (Eds.), *Organizational Ethics, Innovation and Sustainable Innovation* (pp. 126–147). Vilnius: Academic Publishing.

Pusinaite, R. and R. Pucetaite. 2015. *Promoting Sustainable Innovation. Recommendations for Politicians*. Kaunas: Vilnius University, Kaunas Faculty of Humanities.

Quinn, R. 1978. Physical deviance and occupational mistreatment: The short, the fat, and the ugly. Unpublished manuscript, University of Michigan.

Rahman, M., M. Á. Rodríguez-Serrano, and M. Lambkin. 2017. Corporate social responsibility and marketing performance: The moderating role of advertising intensity. *Journal of Advertising Research*, *57*(4): 368–378.

Rajan, R. and R. Ganesan. 2017. A critical analysis of John P. Kotter's Change Management Framework. *Asian Journal of Research in Business Economics and Management*, *7*(7): 181–203.

Rajapathirana, R. P. J. and Y. Hui. 2018. Relationship between innovation capability, innovation type, and firm performance. *Journal of Innovation & Knowledge*, *3*: 44–55.

Raworth, K. 2012. *A Safe and Just Space for Humanity: Can We Live within the Doughnut?* Oxford: Oxfam.

Reave, L. 2005. Spiritual values and practices related to leadership effectiveness. *The Leadership Quarterly*, *16*: 655–687.

Rockström, J., W. Steffen, K. Noone, Å. Persson, F. S. Chapin, E. F. Lambin, T. M. Lenton, M. Scheffer, C. Folke, and H. J. Schellnhuber. 2009. A safe operating space for humanity. *Nature*, *461*: 472–475.

Rodriguez, M. 2014. Innovation, knowledge spillovers and high-tech services in European regions. *Engineering Economics*, *25*(1): 31–39.

Rollins, T. and D. Roberts. 1998. *Work culture, Organisational Performance and Business Success*. London: Quorum Books.

Rose, R. C., N. Kumar, H. Abdullah, and G. Y. Ling. 2008. Organisational culture as a root of performance improvement: Research and recommendations. *Contemporary Management Research*, *4*(1): 43–56.

Rosińska-Bukowska, M. and I. Penc-Pietrzak. 2015. Corporate social responsibility in corporate strategy in the globalised economy. *Administracja i Zarządzanie*, *33*(106): 195–208.
Ross, L., M. Rix, and J. Gold. 2005a. Learning distributed leadership: Part 1. *Industrial and Commercial Training*, *37*: 130–137.
Ross, L., M. Rix, and J. Gold. 2005b. Learning distributed leadership: Part 2. *Industrial and Commercial Training*, *37*: 224–231.
Rothstein, M. G. and R. J. Burke. 2010. *Self-Management and Leadership Development*. Cheltenham: Edward Elgar Publishing.
Rowe, W. G. 2001. Creating wealth in organizations: The role of strategic leadership. *Academy of Management Executive*, *15*(1): 81–91.
Ruderman, M. N., K. Hannum, J. B. Leslie, and J. L. Steed. 2001. Making the connection: Leadership skills and emotional intelligence. *LIA*, *21*(5): 3–7.
Russell, R. F. and A. G. Stone. 2002. A review of servant leadership attributes: Developing a practical model. *Leadership & Organisation Development Journal*, *23*: 145–157.
Ryan, J. C. and S. A. Tipu. 2013. Leadership effects on innovation propensity: A two-factor full range leadership model, *Journal of Business Research*, *66*: 2116–2129.
Saaksjarvi, M. 2003. Consumer adoption of technological innovations. *European Journal of Innovation Management*, *6*(2):90–100.
Saatcioglu, Y and N. Özmen. 2010. Analyzing the barriers encountered in innovation process through interpretive structural modelling: Evidence From Turkey. *Yönetim ve Ekonomi*, *17*(2): 207–225.
Saeed, M. and F. Arshad. 2012. Corporate social responsibility as a source of competitive advantage: The mediating role of social capital and reputational capital. *Database Marketing & Customer Strategy Management*, *19*(4): 219–232.
Saeidi, S. P., S. Sofian, P. Saeidi, S. P. Saeidi, and S. A. Saaeidi. 2015. How does corporate social responsibility contribute to firm financial performance? The mediating role of competitive advantage, reputation, and customer satisfaction. *Journal of Business Research*, *68*(2): 341–350.
Salaman, G., J. Storey, and J. Billsberry. 2005. *Strategic Human Resource Management: Theory and Practice*. 2nd Edition. Thousand Oaks, CA: Sage Publications Ltd.
Samad, S. 2012. The influence of innovation and transformational leadership on organisational performance. *Procedia Social and Behavioural Sciences*, *57*: 486–493.
Sapiegiene, L., V. Jukneviciene, and S. Stoskus. 2009. The process of innovation implementation: A case study of manufacturing companies in Šiauliai city. *Economics and Management: Topicalities and Perspectives*, *2*(15): 237–249.
Sarasvathy, S. D. 2005. *What Makes Entrepreneurs Entrepreneurial?* Virginia: Darden Business Publishing, Social Science Research Network, UVA-ENT-0065.
Schneider, E. D. and J. K. Kay. 1994. Life as a manifestation of the second law of thermodynamics. *Mathematical and Computer Modelling*, *19*(6–8): 25–48.
Schoorl, E. 2012. *Jean-Baptiste Say: Revolutionary, Entrepreneur, Economist.* Routledge Studies in the History of Economics. London: Routledge.
Schumpeter, J. A. 1934. *The Theory of Economic Development: An Inquiry into Profits, Capital, Credits, Interest, and the Business Cycle*. Piscataway, NJ: Transaction Publishers.
Seibt, R., S. Spitzer, M. Blank, and K. Scheuch. 2009. Predictors of work ability in occupations with psychological stress. *Journal of Public Health*, *17*(1): 9–18.
Seitanidi, M. M. and A. Crane. 2009. Implementing CSR through partnerships: Understanding the selection, design and institutionalisation of nonprofit-business partnerships. *Journal of Business Ethics*, *85*(2): 413–429.
Shah, S. S. H., A. Jabran, R. J. Ahsan, W. Sidra, E. Wasiq, F. Maira, and S. K. Sherazi. 2012. Impact of stress on employee's performance: A study on teachers of private colleges of Rawalpindi. *Asian Journal of Business Management*, *4*(2): 101–104.
Shane, S. 2010. *Born Entrepreneurs, Born Leaders: How Your Genes Affect Your Work Life*. New York: Oxford University Press.

Shapero, A. 1975 (November). The displaced, uncomfortable entrepreneur. *Psychology Today*, *9*: 83–133.

Shehri, M. A., P. McLaughlin, A. Al-Ashaab, and R. Hamad. 2017. The impact of organisational culture on employee engagement in Saudi banks. *Journal of Human Resources Management Research*, doi: 10.5171/2017.761672.

Shenhar, A. J. and Dvir, D. 2007. *Reinventing Project Management: The Diamond Approach to Successful Growth and Innovation*. Boston, MA: Harvard Business School Press.

Simanskiene, L. and J. Pauzuoliene. 2011. Perception of the concept of sustainable development in organizations. *Management Theory and Studies for Rural Business and Infrastructure Development*, *2*(26), Klaipėda: Klaipėda University, Klaipėda State College.

Simanskiene, L. and A. Petrulis. 2014. Coherence and its benefits for organizations. *Regional Formation and Development Studies*, *1*(11): 221–229.

Simanskiene, L. and E. Zuperkiene. 2013. *Sustainable Leadership*. Klaipėda: Klaipėda University Press.

Sirgy, M. J., D. Efraty, P. Siegel, and D. Lee. 2001. A new measure of quality of work life (QoWL) based on need satisfaction and spillover theory. *Social Indicators Research*, *55*: 241–302.

Siriwardhane, P. and D. Taylor. 2014. Stakeholder prioritization by mayors and CEOs in infrastructure asset decisions. *Journal of Accounting and Organisational Change*, *10*(3): 1–31.

Skarzynski, P. and R. Gibson. 2008. *Innovation to the Core: A Blueprint for Transforming the Way Your Company Innovates*. Boston, MA: Harvard Business School Press.

Spangenberg, J. H. 2002a. Institutions for sustainable development: Indicators for performance assessment. In IIG, Institut Internacional de Governabilitat (Eds.), *Governance for Sustainable Development* (pp. 133–162). Barcelona: IIG.

Spangenberg, J. H. 2002b. Institutional sustainability indicators: An analysis of the institutions in Agenda 21 and a draft set of indicators for monitoring their effectivity. *Sustainable Development*, *10*(2):103–115.

Staniskis, J., V. Arbaciauskas, Z. Stasiskiene, and V. Varzinskas. 2007. Research work "Analysis and Proposals of Sustainable Industrial Development in Lithuania." Ministry of Economy of the Republic of Lithuania.

Staniskis, J. and Z. Stasiskiene. 2008. Cleaner production innovation development and deployment system (APINI-SPIN). *Environmental Research, Engineering and Management*, *4*(42): 54–59.

Stankevice, I. 2014. Innovation strategies: The patterns across countries, industries and firm competitiveness. *European Scientific Journal*, February 2014, Special edition vol. 1.

Steffen, W., K. Richardson, J. Rockström, S. E. Cornell, I. Fetzer, E. M. Bennett, R. Biggs, S. R. Carpenter, W. de Vries, and C. A. de Wit. 2015. Planetary boundaries: Guiding human development on a changing planet. *Science*, *347*: 736, 1259855.

Steven, W. P. 2000. Organisational culture and its relationship between job tension in measuring outcomes among business executives. *Journal of Management Development*, *19*(1): 32–49.

Stevens, B. 2008. Corporate ethical codes. Effective instruments for influencing behaviour. *Journal of Business Ethics*, *78*: 601–609.

Stevenson, H. H. and C. J. A. Jarillo. 1990. Paradigm of entrepreneurship: Entrepreneurial management. *Strategic Management Journal*, *11*: 17–27.

Stoneman, P. and G. Battisti. 2010. The diffusion of new technology. *Handbook of the Economics of Innovation*, *2*: 733–760.

Stossel, J. 2006. *Myths, Lies, and Downright Stupidity: Get Out the Shovel. Why Everything You Know Is Wrong*. New York: Hyperion.

Stripeikis, O. 2007. Small business in Lithuania: Challenges and development opportunities. *Organizational Management: Systematic Research*, *44*: 142–143.

Suri, V. and D. Chapman. 1998. Economic growth, trade and energy: Implications for the environmental Kuznets curve. *Ecological Economics*, *25*: 195–208.
Susniene, D. and A. Jurkauskas. 2009. The concepts of quality of life and happiness—correlation and differences. *Engineering Economics*, *3*: 58–66.
Tapaninen, A. 2008. Characteristics of innovation: A customer-centric view to adoption barriers of the wood pellet heating system. *International Association for Management of Technology*, *2008*: 1–13.
Tepperman, L. and J. Curtis. 2003. Social problems of the future. *Journal of Futures Studies*, *8*(1): 21–38.
Terziovski, M. 2007. *Building Innovation Capability in Organisations: An International Cross-Case Perspective*. Series on Technology Management, Vol. 13. London: Imperial College Press.
Tezcan Uysal, J. H., S. Aydemir, and E. Genç. 2017. Maslow's hierarchy of needs in the 21st century: The examination of vocational differences. In H. Arapgirlioğlu, R. L. Elliott, E. Turgeon, and A. Atik (Eds.), *Researches on Science and Art in 21st Century Turkey* (Volume 1, pp.211–227). Ankara: Gece Kitaplığı.
Thor, S. and C. Johnson. 2011. Leadership, emotional intelligence, and work engagement: A literature review. *Leadership & Organisational Management Journal*, *3*: 1–66.
Tidd. J., J. Bessant, and K. Pavitt. 1997. *Managing Innovation: Integrating Technological, Market and Organisational Change*. London: Wiley.
Tilly, F. and W. Young. 2009. Sustainability entrepreneurs: Could they be the true wealth generators of the future? *Greener Management International*, *55*(5): 79–92.
Timmons, J. A. 1989. *The Entrepreneurial Mind*. Baltimore, MD: Brick House Publication Company.
Ting, S. and P. Scisco. 2006. *The CCL Handbook of Coaching: A Guide for the Leader Coach*. San Francisco, CA: Jossey-Bass.
Turner, T. and W. W. Pennington III. 2015. Organizational networks and the process of corporate entrepreneurship: How the motivation, opportunity, and ability to act affect firm knowledge, learning, and innovation. *Small Business Economics*, *45*(2): 447–463.
Tyson, L. D. 2004. The economic times. http://articles.economictimes.indiatimes.com/2004-05-02/news/27386154_1_socialentrepreneurs-social-enterprises-social-services.
UNCSD. 2001. *Indicators of Sustainable Development: Guidelines and Methodologies*. New York: United Nations.
United Nations. 2002. *National Implementation of Agenda 21: A Report*. New York: Department of Economic and Social Affairs, Division for Sustainable Development, National Information Analysis Unit, United Nations.
United Nations. 2009. *Eco-Efficiency Indicators: Measuring Resource-Use Efficiency and the Impact of Economic Activities on the Environment*. New York: United Nations.
United Nations. 2014. *The Road to Dignity by 2030: Ending Poverty, Transforming All Lives and Protecting the Planet*. New York: United Nations.
United Nations. 2018. *Working Together: Integration, Institutions and the Sustainable Development Goals, World Public Sector Report 2018*. New York: Division for Public Administration and Development Management, Department of Economic and Social Affairs, (DPADM), April.
United Nations Development Programme. 2010. *Human Development Report*. http://hdr.undp.org/en/content/human-development-report-2010 (accessed July 23, 2019).
United Nations Development Programme. 2011. *Human Development Report*. http://hdr.undp.org/en/content/human-development-report-2011 (accessed July 23, 2019).
United Nations General Assembly. 2012. *The Future We Want* (A/RES/66/288). New York: United Nations.
University of Leicester. 2006. Psychologist produces the first-ever "World Map Of Happiness." *Science Daily*, November 14. www.sciencedaily.com/releases/2006/11/061113093726.htm

Vanhamme, J. and B. Grobben. 2009. Too good to be true! The effectiveness of CSR history in countering negative publicity. *Journal of Business Ethics*, *85*: 273–283.
Varkulevicius, R. and K. Naudzius. 2005. *Methodology for Developing an Innovative Business Plan.* Vilnius: Rosma.
Vijeikiene, B. and J. Vijeikis. 2000. *Basics of Teamwork*. Vilnius: Rosma.
Vijeikis, J. 2011. *Innovation Management*. Vilnius: Technologija.
Waclawski, J. 1998. The relationship between individual personality orientation and executive leadership behaviour. *Journal of Occupational and Organisational Psychology*, *71*: 99–126.
Waddock, S. and B. K. Googins. 2011. The paradoxes of communicating corporate social responsibility. In Ø. Ihlen, J. L. Bartlett, and S. May (Eds.), *The Handbook of Communication and Corporate Social Responsibility* (pp. 23–43). New York: Wiley.
Wall, G. 1990. Exergy conversion in Japanese society, *Energy*, *15*(5): 435–444.
Wall, G. 1997. Exergy use in the Swedish society 1994. International Conference on Thermodynamic Analysis and Improvement of Energy Systems, TAIES'97, Beijing, China, Jun 10–13, 1997 Beijing World.
Weinberger, L. A. 2003. An examination of the relationship between emotional intelligence, leadership style, and perceived leadership effectiveness. PhD dissertation, University of Minnesota, MN (Publication No. AAT 3113218).
Weiss, P. 2004. *The Three Levels of Coaching*. San Francisco, CA: An Appropriate Response.
Westwood, R. and A. Chan. 2002. Headship and leadership. In R. I. Westwood (Ed.), *Organisational Behaviour: Southeast Asian Perspectives* (3rd Edition, pp. 118–143). Hong Kong: Longman.
White, M. D. 2014. The problems with measuring and using happiness for policy purposes. Mercatus Working Paper, Mercatus Center at George Mason University, Arlington, VA, December 2014. http://mercatus.org/publication/problems-measuring-and-using-happiness-policy-purposes (accessed July 23, 2019).
Whitmore, J. 2002. *Coaching for Performance: Growing People, Performance and Purpose*. 3rd Edition. London: Nicholas Brealey Publishing.
Wickham, P. A. 2006. *Strategic Entrepreneurship*. Essex: Prentice Hall.
Wong, C. and K. S. Law. 2002. The effects of leader and follower emotional intelligence on performance and attitude: An exploratory study. *The Leadership Quarterly*, *13*: 243–274.
World Business Council for Sustainable Development. 2014. *WBCSD Submission of Climate Change Business Solutions to the United Nations Climate Summit*. Geneva: WBCSD.
World Commission on Environment and Development. 1987. *Our Common Future*. Oxford, New York: Oxford University Press.
World Economic Forum. 2000. *Pilot Environmental Sustainability Index. An Initiative of the Global Leaders for Tomorrow Environmental Task Force.* Davos: World Development Forum.
World Economic Forum. 2014. *Towards the Circular Economy: Accelerating the Scale-Up Across Global Supply Chains*. Prepared in collaboration with the Ellen MacArthur Foundation and McKinsey & Company. Geneva: World Economic Forum.
World Health Organization. 2014. *Preventing Suicide: A Global Imperative*. Geneva: World Health Organization.
Wynarczyk, P. 2013. Open innovation in SMEs: A dynamic approach to modern entrepreneurship in the twenty-first century. *Journal of Small Business and Enterprise Development*, *20*(2): 258–278.
Yammarino, F. J. 1999. CEO charismatic leadership: Levels-of-management and levels-of analysis effects. *Academy of Management Review*, *24*: 266–286.
Yoon, E. and S. Tello. 2009. Drivers of sustainable innovation: Exploratory views and corporate strategies. *Seoul Journal of Business*, *15*(2): 85–115.

York, J. G. and S. Venkataraman. 2010. The entrepreneur–environment nexus: Uncertainty, innovation, and allocation. *Journal of Business Venturing*, *25*(5): 449–463.

Young, A. J., Y. H. Yom, and J. S. Ruggiero. 2011. Organisational culture, quality of work life, and organisational effectiveness in Korean university hospitals. *Journal of Transcultural Nursing*, *22*(1): 22–30.

Zakarevičius, P. 2004. *Management: Genesis, Present, Trends*. Kaunas: VMU Publishing House.

Zakas, N. C. 2005. The eye of the beholder: Appearance discrimination in the workplace. Master's thesis, Endicott College, MA.

Zeidner, M., G. Matthews, and R. Roberts. 2004. Emotional intelligence in the workplace: A critical review. *Applied Psychology: An International Review*, *53*: 371–399.

Židonis, Ž. 2008. Entrepreneurship promotion policy in Lithuania: Productive, non-productive and destructive entrepreneurship. *Public Policy and Administration*, *26*(1). Vilnius: MRU.

Zizlavsky, O. 2013. Past, present and future of the innovation process. *International Journal of Engineering Business Management*, *5*: 1–8.

Zohar, D. and I. Marshall. 2000. *SQ: Spiritual Intelligence. The Ultimate Intelligence*. London: Bloomsbury.

Zohar, D. and I. Marshall. 2004. *Spiritual Capital. Wealth We Can Live By*. San Francisco, CA: Berrett-Koehler.

Zonooz, B. H., v. Farzam, M. Satarifar, and L. Bakhshi. 2011. The relationship between knowledge transfer and competitiveness in "SMES" with emphasis on absorptive capacity and combinative capabilities. *International Business and Management*, *2*(1): 59–85.

# Index

## F



## G



## H



## I

## K



## L



## M

## N



## O



## P



## Q

## T

## U



## V



## W